Digging in Assyria

Edited by
Shirley Glubok

Abridged and adapted from
NINEVEH AND ITS REMAINS
by AUSTEN HENRY LAYARD

Foreword by Prudence Oliver Harper, Associate Curator
of Ancient Near Eastern Art, The Metropolitan Museum of Art

Designed by Gerard Nook

THE MACMILLAN COMPANY
COLLIER-MACMILLAN LTD., LONDON

Other books by Shirley Glubok:

THE ART OF ANCIENT EGYPT
THE ART OF LANDS IN THE BIBLE
THE ART OF ANCIENT GREECE
THE ART OF THE NORTH AMERICAN INDIAN
THE ART OF THE ESKIMO
THE ART OF ANCIENT ROME
THE ART OF AFRICA
ART AND ARCHAEOLOGY
THE ART OF ANCIENT PERU
THE ART OF THE ETRUSCANS
THE ART OF ANCIENT MEXICO
KNIGHTS IN ARMOR
THE ART OF INDIA
THE ART OF JAPAN
THE ART OF COLONIAL AMERICA
THE FALL OF THE AZTECS
THE FALL OF THE INCAS
DISCOVERING TUT-ANKH-AMEN'S TOMB
DISCOVERING THE ROYAL TOMBS AT UR
HOME AND CHILD LIFE IN COLONIAL DAYS

The Macmillan Company, 866 Third Avenue, New York, New York 10022
Collier-Macmillan Canada Ltd., Toronto, Ontario
Library of Congress Catalog Card Number: 79-103679
First Printing

Pages 2-3 and 126-127 illustration: *The Metropolitan Museum of Art, Gift of John D. Rockefeller, Jr., 1932*

Page 4 illustration (detail): *The Metropolitan Museum of Art, Gift of John D. Rockefeller, Jr., 1932*

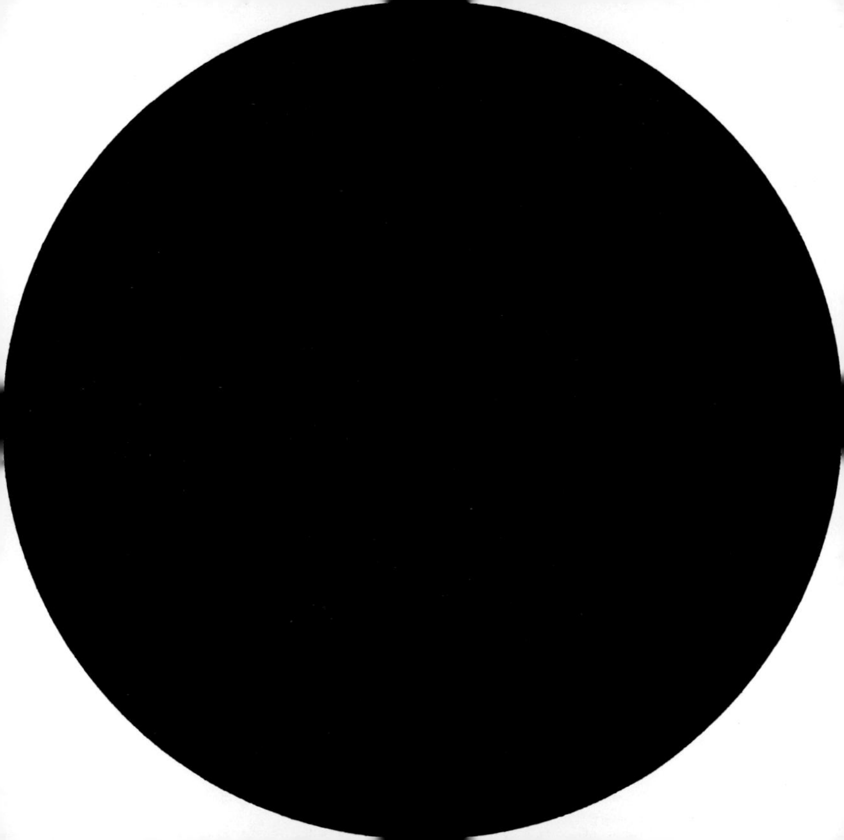

Foreword

In 1845 Austen Henry Layard began excavating a mound called Nimrud in northern Iraq. In the course of the next two years he unearthed there, and at the neighboring site of Kuyunjik, the remains of a people about whom almost nothing was then known. . . . They were the Assyrians. In this volume Layard describes his discovery of their palaces and temples, and tells of the trials and tribulations as well as the joys of archaeological excavation. The differences between this record of a century ago, and the story of present-day expeditions in the same area are vast. The Turkish officials who sometimes helped—but more often hindered—Layard's work are gone. The job of supervising the workmen, of drawing, recording, surveying and preserving whatever is discovered is now done by a team of scholars. Layard did all this alone and also arranged to have the immense sculptures he found shipped back to England.

The history of the Assyrian empire, which existed in the north of Mesopotamia from the ninth to the seventh century B.C., is largely one of wars and conquests: north into Anatolia, south into Babylonia, east into Iran, and chiefly west to conquer and gain a firm hold on the vital trade routes leading to the Mediterranean Sea. Year after year the Assyrian kings set out on expeditions to preserve this empire, while at home internal conflicts came with the accession of each new ruler to the throne. We read of these events not only in the annals of the Assyrian kings, but also in the historical books of the Old Testament, which created the first real interest, a century ago, in the exploration of Assyria and the discovery of the Assyrian cities of Nineveh and Calah. The final collapse of Assyrian rule and the destruction of both Nimrud and Nineveh occurred in 612 B.C. Assyria fell, and the Babylonians to the south rose to power in Mesopotamia.

Austen Henry Layard began his work at Nimrud, now known to have been Calah, one of the ancient capital cities of Assyria. He writes that almost immediately he came upon the remains of two palaces with huge limestone sculptures and reliefs decorating the entrances and inside walls. He describes these as they appeared, noting the changes in the form of the dress, and in the arms and equipment illustrated in the scenes. Layard correctly interpreted these differences in detail as the reflection of the passage in time from the reign of one king to another. The cuneiform inscriptions on the reliefs gave him the names and genealogies of the kings of Assyria, although in Layard's time it was still not possible to decipher these inscriptions with complete accuracy.

At Nimrud the principal discovery was the Northwest Palace of Ashurnasirpal II (883–859 B.C.).

The reliefs used to decorate the walls in the Southwest Palace had been taken from the Northwest Palace and from another palace in the center of the mound by a later king, Esarhaddon (680–669 B.C.), a fact which Layard recognized and recorded.

The reliefs from the Center Palace were actually found by Layard, stacked up and ready for removal to the Southwest Palace. Esarhaddon died, however, before the Southwest Palace could be completed.

The job of excavating the central part of the mound still remains. Later expeditions to Nimrud left it largely untouched because of the accumulation of debris above the palace.

In the southeast area Layard discovered what later excavations proved to be another palace as well as a large complex of buildings including the temple of Nabu, the god of science and learning.

Layard called the great stepped tower at the north corner of the mound a pyramid and believed it to be a tomb. Actually it is a ziggurat, a type of sacred building which occurs at many Mesopotamian sites. Its exact function is unknown.

At Kuyunjik, close to the modern town of Mosul, the excavation was more difficult than at Nimrud. The depth of the earth above the ruins was immense and only in Layard's second season of excavation (recorded in another book, *Nineveh and Babylon*) was the immense palace built by Sennacherib, the grandson of Esarhaddon, (704–681 B.C.) cleared to any great extent.

After Layard finally departed from Assyria, and gave up his career as an archaeologist to enter the field of diplomacy, excavations continued sporadically at the sites he describes in this volume. The British went on for a time at both Nimrud and Nineveh. Then in 1953, after a gap of almost a century, an expedition led by Sir M. E. L. Mallowan returned to Nimrud. Mallowan discovered additional rooms in the palaces and temples already excavated by Layard, as well as some entirely new buildings on the citadel, in the town below and on the city walls. These he described in two large volumes called, in memory of Layard, *Nimrud and Its Remains.*

As a result of this work, begun by Layard and carried on by other archaeologists, more is known of the history of Assyria in the first millennium B.C.; of Assyrian customs, crafts and life, both in war and peace, than of any of the earlier civilizations in Mesopotamia. The stone reliefs themselves supply much of this information and are a type of record which is completely lacking for the earlier periods of Near Eastern history. Large quantities of wall paintings must originally have existed but they are so fragile that they have not, for the most part, survived the passage of centuries.

Nimrud was not the ancient Assyrian capital of Nineveh or even part of Nineveh as Layard thought but rather another city altogether, Calah. The true location of Nineveh was the mound of Kuyunjik. A third Assyrian capital was at Khorsabad. The Assyrians called it Dur Sharruken, literally 'the city of Sargon,' after the king (721–705 B.C.) who built his palace there. Assur, the most ancient capital city of the Assyrians, lay under the ruins at Kalah Sherqat, a site to the south that Layard visited on the Tigris below Nimrud.

If Layard's descriptions of tunneling through the mounds of Nimrud and Kuyunjik, following the lines of stone sculptures, sound horrifyingly unprofessional and hopelessly haphazard to the modern archaeologist, it is to be remembered that this man supervised alone, with relatively simple equipment, the discovery of an unknown civilization of tremendous wealth and of immense material remains. His achievements in Assyria, which he records in part here, are a remarkable chapter in the early history of archaeology in the Near East.

Prudence Oliver Harper
Associate Curator of Ancient
Near Eastern Art,
The Metropolitan Museum of Art

New York City
May, 1970

CHRONOLOGY

Following are the names and dates of the kings during whose reigns the buildings and monuments found by Layard were chiefly made.

Nimrud—ancient Calah

Northwest Palace—Ashurnasirpal II 883-859 B.C.
Southwest Palace—Esarhaddon 680-669 B.C.
Center Palace (possibly two palaces are here)—
Shalmaneser III 858-824 B.C.
Tiglath-Pileser III 744-727 B.C.
Central area—Obelisk of Shalmaneser III 858-824 B.C.
Southeast area
Palace—probably Sargon 721-705 B.C.
Temple of Nabu—probably Sargon 721-705 B.C.

Kuyunjik—ancient Nineveh

Palace of Sennacherib 704-681 B.C.

Introduction

During the latter half of the nineteenth century, the name Austen Henry Layard was known throughout England. A pioneer of Assyrian archaeology, Layard wrote of his excavations and travels in a style that fascinated the general public. His *Nineveh and Its Remains*, published in 1849, met with instant popularity. Eager readers thronged to the British Museum to admire the monumental sculptures that he had unearthed in Assyria.

Layard was born in Paris, in 1817, of English parents. During his boyhood, his father's poor health required constant changes of climate, so the family lived for extensive periods in Switzerland and Italy. In Florence, young Layard acquired a taste for the fine arts.

The boy read many of the books in his father's library, including several on the mysterious and fabled East. The romance of the *Arabian Nights* captured his imagination and he began to dream of travel and adventure.

At the age of sixteen Layard was sent to London to study law in an uncle's office. He remained there for nearly six years, gradually becoming aware that he was not suited to the profession. He turned his interests more and more to the Orient.

In 1839, Layard was offered a position in the civil service in Ceylon. Thrilled with the opportunity to leave London and to travel in the East, he planned a journey across Europe, Central Asia and India. In preparation for the trip he acquired some knowledge of medicine and learned how to plot routes. He had already studied Arabic, and he knew a little Persian.

In July, 1839, Layard set out on the difficult overland route with a single companion. But he was not to reach Ceylon. After traveling for many months in the Middle East and Central Asia, in regions which had rarely been seen by Europeans, he accepted a position in the British foreign service in Constantinople (now Istanbul) in 1842. He remained there for two years. While in Constantinople, he met Paul Emile Botta, the French diplomat and archaeologist who was the first to uncover the remains of the Assyrian civilization. At the mound of Khorsabad, Botta had excavated the ruins of an immense palace. Layard was inspired by Botta to search for other remains of the ancient Assyrian empire.

In preparation for his excavations, Layard learned various Semitic languages which he felt might help him to decipher the cuneiform inscriptions—the wedge-shaped characters which appear on Assyrian stone sculpture.

The story of the first two years of Layard's excavations is told in his two-volume work, *Nineveh and Its Remains*. I have abridged these volumes here, placing primary emphasis on Layard's archaeological achievements. I have retained his original prose with all its enthusiasm and color, and have kept his spellings of place names: Nimroud (modern spelling Nimrud) and Kouyunjik (modern spelling Kuyunjik), for example. A few common nouns and verbs have been changed from the British to the American spellings and an occasional explanation for a technical or foreign word has been inserted.

Photography had not been developed at the time of Layard's Assyrian excavations. The archaeologist himself made the drawings which were later used to illustrate his writings. Several of them are reproduced in this book.

I am grateful to Prudence Oliver Harper for her interest and encouragement in this project. My thanks also to Vaughn Crawford, Curator of Ancient Near Eastern Art at The Metropolitan Museum of Art, for his advice on language problems; to R. D. Barnett, Keeper of West Asiatic Antiquities at the British Museum, for his guidance in picture research; and to C. B. F. Walker, Research Assistant at the British Museum, for help in obtaining photographs.

For further reading, I recommend Layard's other works, particularly *Early Adventures in Persia, Susiana, and Babylonia* (London, 1887) and, for the story of his subsequent excavations, *Discoveries Among the Ruins of Nineveh and Babylon* (London, 1853). C. J. Gadd's *Stones of Assyria* (London, 1936) is a comprehensive account of the sculptured reliefs found in Assyria by Layard and other archaeologists. *Assyrian Palace Reliefs* (London, 1960) and *A Catalogue of the Nimrud Ivories in the British Museum* (London, 1957) both by R. D. Barnett and *The Great King, King of Assyria* by Edith Porada (The Metropolitan Museum of Art, New York, 1946) are also recommended. *Road to Nineveh* by Nora Benjamin Kubie (New York, 1964) and *Layard of Nineveh* by Gordon Waterfield (New York, 1968) add to the picture of Layard. *Foundations in the Dust* by Seton Lloyd (London, 1947) gives a complete account of excavations in Mesopotamia, and Sir M. E. L. Mallowan's *Nimrud and Its Remains* (New York, 1966) continues the story of archaeology in Assyria and completes the record of Layard's accomplishments.

New York City
April, 1970

The Metropolitan Museum of Art,
Gift of John D. Rockefeller, Jr., 1932 (detail)

Chapter 1

■ During the autumn of 1839 and winter of 1840, I had been wandering through Asia Minor and Syria, accompanied by one no less curious and enthusiastic than myself [Edward Mitford]. We were both careless of comfort and unmindful of danger. We rode alone; our arms were our only protection. A valise behind our saddles was our wardrobe, and we tended our own horses, except when relieved from the duty by the hospitable inhabitants of a Turcoman village or an Arab tent.

I had traversed Asia Minor and Syria, visiting the ancient seats of civilization and the spots which religion has made holy. I now felt an irresistible desire to penetrate to the regions beyond the Euphrates River, to which history and tradition point as the birthplace of the wisdom of the West.

A deep mystery hangs over Assyria, Babylonia and Chaldea. With these names are linked great nations and great cities dimly shadowed forth in history; mighty ruins, in the midst of deserts, defying by their very desolation and lack of definite form the description of the traveler; the remnants of mighty races still roving over the land; fulfillment of prophecies; the plains to which the Jew and the Gentile alike look as the cradle of their race. Without treading on the remains of Nineveh and Babylon our pilgrimage was incomplete.

My companion and I ran daily risks and more than once found ourselves in the midst of foraging parties and of tents which, an hour before, had been pillaged by the wandering bands of Arabs.

During a short stay in Mosul we visited the great ruins on the east bank of the Euphrates, which have been generally believed to be the remains of Nineveh. (These ruins include the great mounds of Kouyunjik and Nebbi Yunus.) We rode also into the desert and explored the mound of Kalah Sherghat, a vast ruin on the Tigris River, about fifty miles below its junction with the [River] Zab.

Kalah Sherghat was an Assyrian ruin: a vast, shapeless mass, now covered with grass and showing scarcely any traces of the work of man except where the winter rains had formed ravines down its almost perpendicular sides and had thus laid open its contents. A few fragments of pottery and inscribed bricks, discovered after a careful search among the rubbish which had accumulated round the base of the great mound, served to prove that it

The Metropolitan Museum of Art, Gift of John D. Rockefeller, Jr., 1932 (detail)

owed its construction to the people who had founded the city of which Nimroud is the remains. There was a tradition current among the Arabs that strange figures carved in black stone still existed among the ruins; but we searched for them in vain.

It was not until the summer of 1842 that I again passed through Mosul. On my arrival I found that Paul Emile Botta had been named French Consul there and had already commenced excavations in the large mound on the opposite side of the river, called in the country Kouyunjik. To him is due the honor of having found the first Assyrian monument. The small party employed by M. Botta were at work on Kouyunjik when a peasant from a distant village chanced to visit the spot. Seeing that any fragments of brick and alabaster uncovered by the workmen were carefully preserved, he asked the reason of this strange proceeding. On being informed that they were in search of sculptured stones, he advised them to try the mound on which his village was built, and in which, he declared, many such things as they wanted had been exposed on digging for the foundations of new houses. M. Botta, having been frequently deceived by similar stories, was not at first inclined to follow the peasant's advice but subsequently sent an agent and one or two workmen to the place. After a little opposition from the inhabitants, they were permitted to sink a well in the mound; and at a small distance from the surface they came to the top of a wall which, on digging deeper, they found to be constructed of sculptured slabs of gypsum. M. Botta, on receiving information of this discovery, went at once to the village, which was called Khorsabad. He directed a wider trench to be formed and to be carried in the direction of the wall. He soon found that he had opened a chamber which was connected with others and constructed of slabs of gypsum covered with sculptured representations of battles, sieges and similar events. His wonder may easily be imagined. A new history had been suddenly opened to him—the records of an unknown people were before him. It was evident that the monument appertained to a very ancient and very civilized people; and it was natural from its position to refer it to the inhabitants of Nineveh, a city which, although it could not have occupied a site so distant from the Tigris River, must have been in the vicinity of the place. M. Botta had discovered an Assyrian edifice, the first, probably, which had been exposed to the view of man since the fall of the Assyrian empire.

Sir Austen Henry Layard,
British Museum

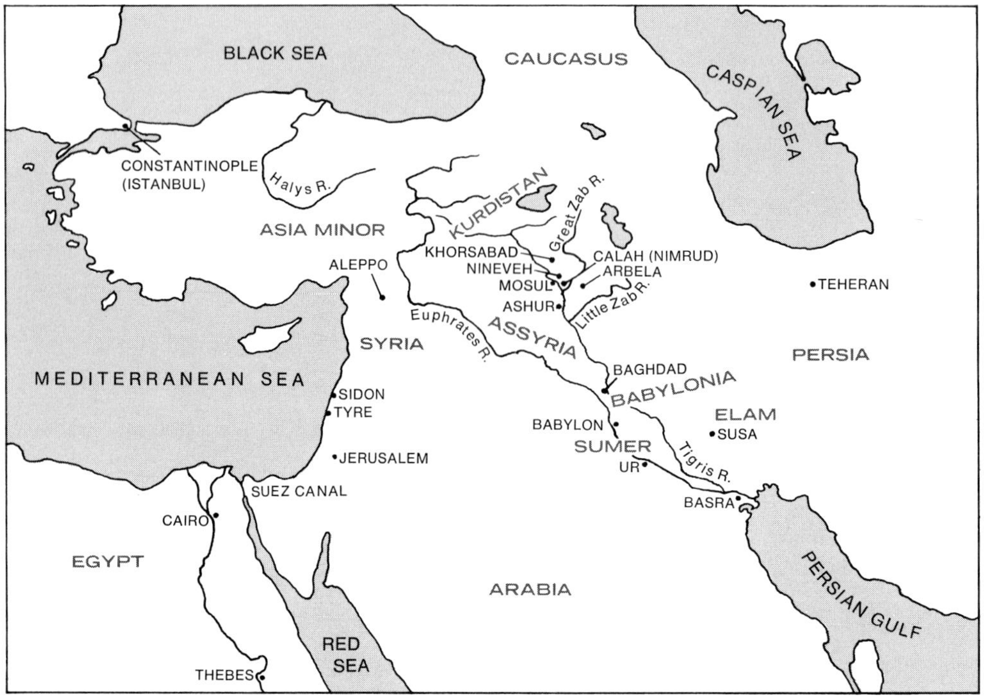

M. Botta was not long in perceiving that the building, which had been thus partly excavated, unfortunately owed its destruction to fire, and that the gypsum slabs, reduced to lime, were rapidly falling to pieces on exposure to the air. No precaution could arrest this rapid decay, and it was to be feared that this wonderful monument had only been uncovered to complete its ruin. The records of victories and triumph, which had long attested the power and swelled the pride of the Assyrian kings and had resisted the ravages of ages, were now passing away forever. Almost all that was first discovered thus speedily disappeared; and the same fate has befallen nearly everything subsequently found at Khorsabad. The researches of M. Botta were not extended beyond Khorsabad; and, having secured many fine specimens of Assyrian sculpture for his country, he returned to Europe with a rich collection of inscriptions, the most important result of his discovery.

The success of M. Botta had increased my anxiety to explore the ruins of Assyria. It was evident that Khorsabad could not stand alone. It did not represent ancient Nineveh, nor did it afford us any additional evidence of the site of that city. If the edifice discovered had been one of its palaces, surely other buildings of a vaster and more magnificent character must exist nearer the seat of government, on the banks of the river Tigris.

My thoughts went back to the ruin of Nimroud and to the traditions which attached to it. At last, in the autumn of 1845, Sir Stratford Canning mentioned to me his readiness to incur, for a limited period, the expenses of excavations in Assyria. My preparations were soon completed, and I started from Constantinople by steamer in the middle of October.

Anxious to reach the end of my journey, I crossed the mountains of Pontus and the great steppes of the Usun Yilak as fast as post horses could carry me, descended the highlands into the valley of the Tigris, galloped over the vast plains of Assyria and reached Mosul in twelve days.

Layard with his servant on a previous journey, British Museum

Chapter 2

■ My first step on reaching Mosul was to present my letters to the Governor of the province, Mohammed Pasha. The appearance of his Excellency was not prepossessing, but it matched his temper and conduct. He had one eye and one ear; he was short and fat, deeply marked by the smallpox, uncouth in gestures and harsh in voice. He had revived many old customs and impositions. He particularly insisted on *dish-parassi* (literally, "tooth money"), or a compensation levied upon all villages in which a man of such rank is entertained, for the wear and tear of his teeth in masticating the food he condescends to receive from the inhabitants. The tribes who had been attacked and plundered [by Mohammed Pasha's men] were retaliating upon caravans and travelers or laying waste the cultivated parts of the Pashalic. The villages were deserted, and the roads were little frequented and very insecure.

Such was the Pasha to whom I was introduced by the British Vice-Consul, Mr. Christian Rassam, two days after my arrival. He received me with that civility which a traveler generally expects from a Turkish functionary of high rank. His anxiety to know the object of my journey was evident.

There were many reasons which rendered it necessary that my plans should be concealed until I was ready. From the authorities and the people of the town I could only expect the most decided opposition. On the eighth of November, having secretly procured a few tools and engaged a mason and carrying with me a variety of guns, spears and other formidable weapons, I declared that I was going to hunt wild boars in a neighboring village and floated down the Tigris on a small raft constructed for my journey.

At this time of the year more than five hours are required to descend the Tigris from Mosul to Nimroud. It was sunset before we reached the dam across the river. We landed and walked to the village of Naifa. No light appeared as we approached, nor were we even saluted by the dogs, which usually abound in an Arab village. We had entered a heap of ruins. I was about to return to the raft, upon which we had made up our minds to pass the night, when a glare of a fire lighted up the entrance to a miserable hovel. Through a crevice in the wall I saw an Arab family crouching round a heap of half-extinguished embers. As we entered, all rose and showed some alarm. The man, however, seeing that we were Europeans, bid us welcome. Our host, whose name was Awad, was a sheikh of the Jehesh. His tribe had been

The Metropolitan Museum of Art, Gift of John D. Rockefeller, Jr., 1932 (detail)

plundered by the Pasha and was now scattered in different parts of the country. He had taken refuge in this ruined village. He told us about marauding excursions on this side of the river. The neighborhood, he said, was consequently insecure, and the roads to Mosul almost closed. Awad had learned a little Turkish and was intelligent and active. Seeing at once that he would be useful, I acquainted him with the object of my journey, offering him the prospect of regular employment in the event of the experiment proving successful, and assigning him regular wages as superintendent of the workmen. He had long been acquainted with the ruins and entertained me with traditions connected with them. [He] volunteered to walk, in the middle of the night, to Selamiyah, a village three miles distant, to procure men to assist in the excavations. He returned with six Arabs, who agreed for a small sum to work under my direction.

The lofty cone and broad mound of Nimroud broke like a distant mountain on the morning sky. The eye wandered over a parched and barren waste, across which occasionally swept the whirlwind, dragging with it a cloud of sand. About a mile from us was the small village, a heap of ruins.

Twenty minutes' walk brought us to the principal mound. The absence of all vegetation enabled me to examine the remains with which it was covered. Broken pottery and fragments of bricks, both inscribed with the cuneiform character, were strewed on all sides. The Arabs watched my motions as I wandered to and fro, and observed with surprise the objects I had collected. They joined in the search, and brought me handfuls of rubbish, among which I found with joy the fragment of a bas-relief. The material on which it was carved had been exposed to fire and resembled, in every respect, the burnt gypsum of Khorsabad. Convinced from this discovery that sculptured remains must still exist in some part of the mound, I sought for a place where excavations might be commenced with a prospect of success. Awad led me to a piece of alabaster which appeared above the soil. We could not remove it, and on digging downward, it proved to be the upper part of a large slab. I ordered all the men to work around it, and they shortly uncovered a second slab to which it had been united. Continuing in the same line, we came upon a third; and, in the course of the morning, laid bare ten more, the whole forming a square with one stone missing at the northwest corner. It was evident that the top of a chamber had been discovered and that the gap was its entrance. (This was the chamber marked A in the Northwest Palace.) I now dug down the face of the stones, and an inscription in the cuneiform character was soon exposed to view. Similar inscriptions occupied the center of all the slabs, which were in the best preservation.

Night interrupted our labors. I was well satisfied with their result. It was now evident that buildings of considerable extent existed in the mound and that although some had been destroyed by fire others had escaped conflagration.

On returning to the village, I moved from the crowded hovel in which we had passed the night. With the assistance of Awad, who was no less pleased than myself with our success, we patched up with mud the least-ruined house in the village and restored its falling roof. We contrived at least to exclude, in some measure, the cold night winds and to obtain a little privacy.

Next morning my workmen were increased by five Turcomans from Selamiyah who had been attracted by the prospect of regular wages. I employed half of them in emptying the chamber partly uncovered on the previous day [in the Northwest Palace] and the rest in following the wall at the southwest corner of the mound. Before evening the work of the former was completed, and I found myself in a room built of slabs about eight feet high and varying from six to four feet in breadth, placed upright and closely fitted together.

In the rubbish near the bottom of the chamber I found several ivory ornaments, upon which were traces of gilding. Awad, who had his own suspicions of the object of my search which he could scarcely persuade himself was limited to mere stones, carefully collected all the scattered fragments of gold leaf he could find in the rubbish and, calling me aside in a mysterious and confidential fashion, produced them wrapped up in a piece of dingy paper. "O Bey," said he, "Wallah! Your books are right. Here is the gold, sure enough, and, please God, we shall find it all in a few days. Only don't say anything about it to those Arabs, for they cannot hold their tongues. The matter will come to the ears of the Pasha." The Sheikh was much surprised when I generously presented him with the treasures he had collected and all such as he might hereafter discover.

As my chief aim was to prove the existence, as soon as possible, of sculptures, all my workmen were moved to the southwest corner. I continued the excavations in this part of the mound until the thirteenth [November], still uncovering inscriptions, but finding no sculptures.

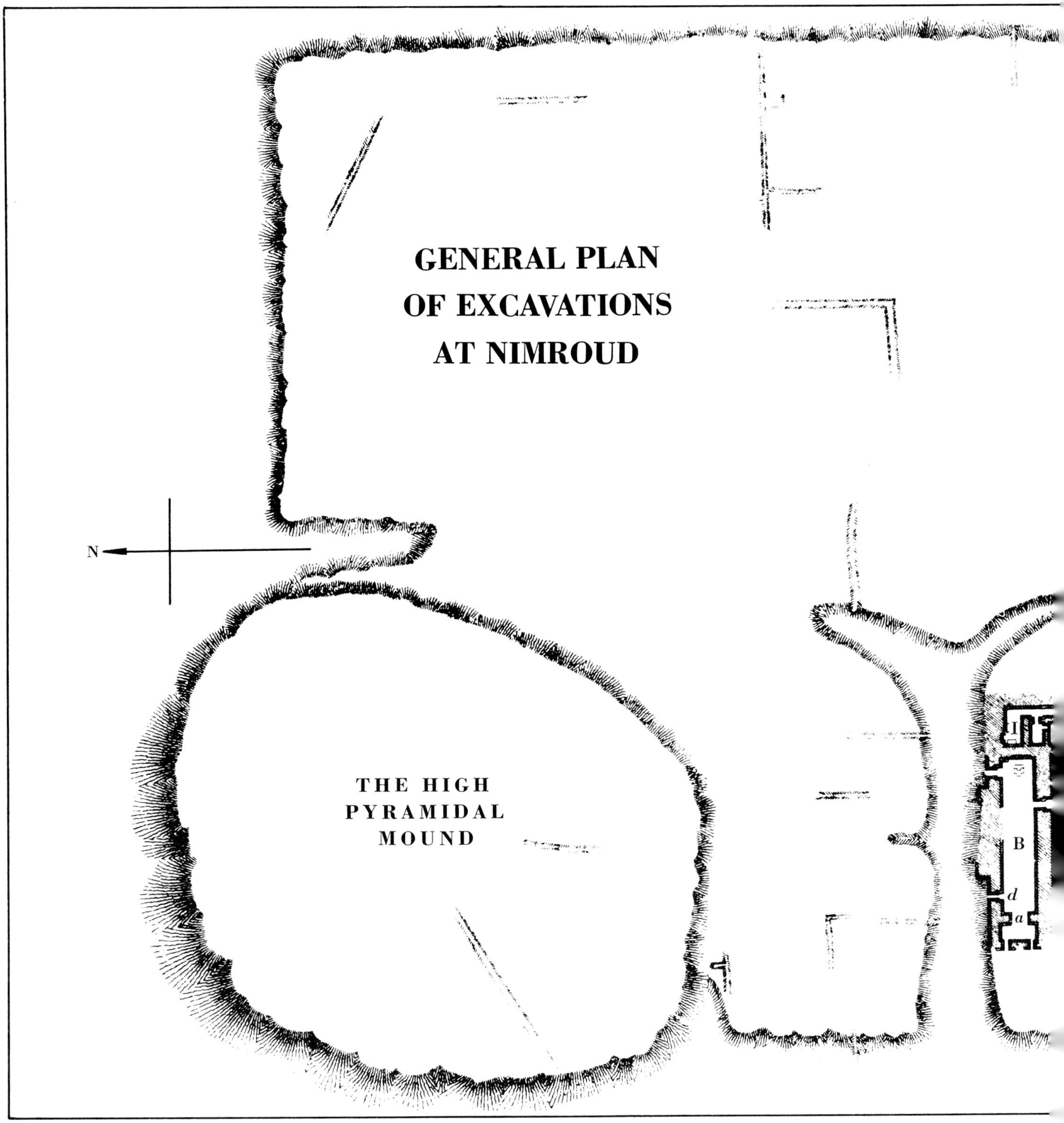

GENERAL PLAN
OF EXCAVATIONS
AT NIMROUD
N
THE HIGH
PYRAMIDAL
MOUND
I
B
d
a

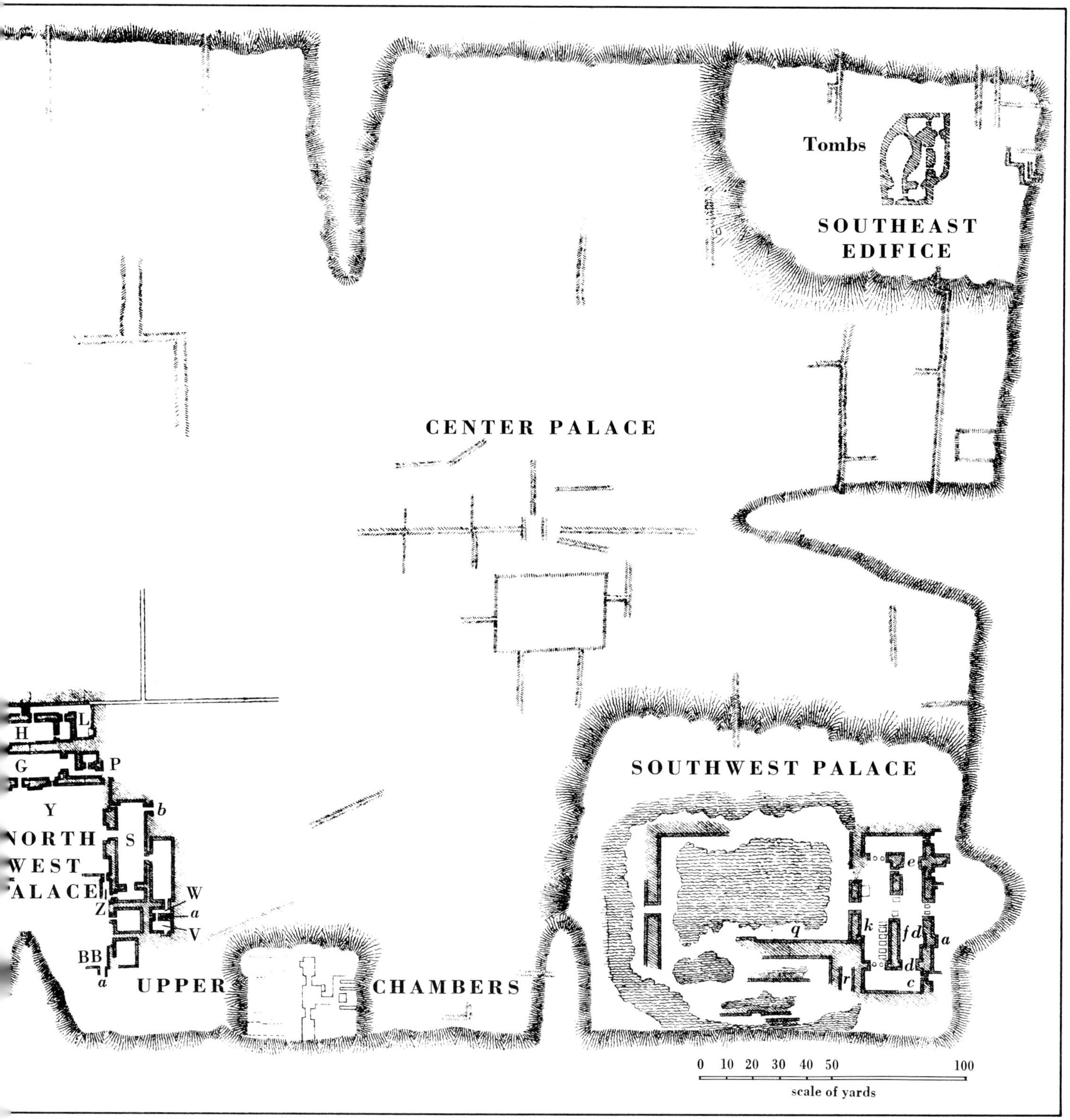
Tombs
SOUTHEAST EDIFICE
CENTER PALACE
SOUTHWEST PALACE
NORTH WEST PALACE
UPPER CHAMBERS
L
H
G
P
Y
S
b
W
a
V
Z
BB
a
e
k
f
d
a
q
r
c
0 10 20 30 40 50 100
scale of yards

Some days having elapsed since my departure from Mosul and the experiment having been now sufficiently tried, it was time to return to the town and acquaint the Pasha, who had, no doubt, already heard of my proceedings, with the object of my researches.

He pretended at first to be ignorant of the excavations at Nimroud; but subsequently thinking that he would convict me of prevarication in my answers to his questions as to the amount of treasure discovered, pulled out of his writing tray a dingy scrap of paper, in which was preserved an almost invisible particle of gold leaf. This, he said, had been brought to him by the commander of the irregular troops stationed at Selamiyah, who had been watching my proceedings. I suggested that he should name an agent to be present as long as I worked at Nimroud, to take charge of all the precious metals that might be discovered. He promised to write on the subject to the chief of the irregulars, but offered no objection to the continuation of my researches.

Reports of the wealth extracted from the ruins had already reached Mosul and had excited the cupidity and jealousy of the cadi [a minor magistrate] and principal inhabitants of the place. Others, who well knew my object and might have spared me any additional interruption without a sacrifice of their national character, were not backward in throwing obstacles in my way and in fanning the prejudices of the authorities and natives of the town. It was evident that I should have to contend against a formidable opposition; but as the Pasha had not, as yet, openly objected to my proceedings, I hired several Nestorian Chaldeans who had left their mountains for the winter to seek employment in Mosul and sent them to Nimroud. At the same time I engaged agents to explore several mounds in the neighborhood of the town, hoping to ascertain the existence of sculptured buildings before steps were taken to interrupt me.

Having finished my arrangements in Mosul, I returned to Nimroud. During my absence, the workmen, under the direction of my cawass [armed constable], had carried the excavations along the back of the wall (*c*, in the Southwest Palace), and had discovered the entrance (*d*). Being anxious to make as much progress as possible, I increased my party to thirty men and distributed them in three sets over the southwest corner of the mound. By opening long trenches at right angles in various directions, we came upon the top of a wall built of slabs with inscriptions. One, however, was reversed, and was covered with characters exceeding in size any I had yet seen. I could not account for its strange position. The edges of this, as well as of all the other slabs in the southwest corner, had been cut away, several letters of the inscriptions being destroyed in order to make the stones fit into the wall. From these facts it was evident that materials taken from another building had been used in the construction of the one we were now exploring. But as yet it could not be ascertained whether the face or the back of the slabs had been uncovered. Neither the plan nor the nature of the edifice could be determined until the heap of rubbish and earth under which it was buried had been removed. The soil, mixed with sun-dried and kiln-burnt bricks, pottery and fragments of alabaster, offered considerable resistance to the tools of the workmen; and, when loosened, had to be removed in baskets and thrown over the edge of the mound. The Chaldeans from the mountains, strong and hardy men, could alone wield the pick; the Arabs were employed in carrying away the earth. The spade could not be used, and there were no other means than those I had adopted to clear away the ruins. A person standing on the mound would see no remains of building until he approached the edge of the trenches into which the workmen descended by steps. Parts of the walls were now exposed to view; but it was impossible to conjecture which course they took or whether the slabs were facing the inside or formed the back of the chamber.

Depredations by Arab tribes continuing in the plains of Nimroud and surrounding country, I deemed it prudent to remove from Naifa, where I had hitherto resided, to Selamiyah.

The Yuz-bashi, or captain of the irregular troops, one Daoud Agha, came to call upon me as soon as I was established in my new quarters. Like most men of his class, he was frank and intelligent. He tendered me his services, entertained me with his adventures and planned hunting expeditions. A few presents secured his adherence, and he proved himself afterward a very useful and faithful ally.

I had now to ride three miles every morning to the mound, and my workmen, who were afraid, on account of the Arabs, to live at Naifa, returned after the day's labor to Selamiyah.

The excavations were still carried on as actively as the means at my disposal would permit. The entrance, *d* (Southwest Palace), had now been completely exposed, and the backs of several slabs of wall *d* had been uncovered. But still no sculptures had been discovered nor could any idea be yet formed of the relative position of the walls. I ordered a trench to be opened obliquely from the entrance *d* into the interior of the mound, presuming that we should ultimately find the opposite side of the chamber, to which, it appeared probable, we had found the passage. After removing a large accumulation of earth mixed with charcoal, charred wood and broken bricks, we reached the top of wall *f* on the afternoon of the twenty-eighth November. In order to ascertain whether we were in the inside of a chamber, the workmen were directed to clear away the earth from both sides of the slabs. The south face was unsculptured, but the first stroke of the pick on the opposite side disclosed the top of a bas-relief. The Arabs were no less excited than myself by the discovery; and notwithstanding a violent shower of rain, working until dark, they completely exposed to view two slabs.

On each slab were two bas-reliefs, separated from one another by a band of inscriptions. The lower bas-relief on the first slab represented the siege of a castle or of a walled city. To the left were two warriors, each holding a circular shield in one hand and a short sword in the other. A tunic, confined at the waist by a girdle and ornamented with a fringe of tassels, descended to the knee; a quiver was suspended at the back and the left arm was passed through the bow, which was thus kept by the side ready for use. They wore pointed helmets. The foremost warrior was ascending a ladder placed against the castle. Three turrets, with angular battlements, rose above walls similarly

British Museum

ornamented. In the first turret were two warriors, the one in the act of discharging an arrow, the other raising a shield and casting a stone at the assailants from whom the besieged were distinguished by their headdress—a simple fillet binding the hair above the temples. Their beards, at the same time, were less carefully arranged. The second turret was occupied by a slinger preparing his sling. In the interval between this turret and the third, and over an arched gateway, was a female figure, known by her long hair descending upon the shoulders in curls. Her right hand was elevated as if in the act of asking for mercy. In the third turret were two more of the besieged, the first discharging an arrow, the second elevating his shield and endeavoring with a torch to burn an instrument resembling a catapult, which had been brought up to the wall by an inclined plane built on a heap of boughs and rubbish. These figures were out of all proportion when compared with the size of the building. A warrior with a pointed helmet, bending on one knee and holding a torch in his right hand, was setting fire to the gate of the castle, while another in full armor was forcing the stones from the foundation with an instrument, probably of iron, resembling a blunt spear. Between them was a wounded man falling headlong from the walls.

The second slab was a cornerstone very much injured, the greater part of the relief having been cut away to reduce it to convenient dimensions. The lower bas-relief represented a singular subject. On the walls of the castle, two stories high and defended by many towers, stood a woman tearing her hair to show her grief. This slab had been exposed to fire like that adjoining and had sustained too much injury to be removed.

As I was meditating in the evening over my discovery, Daoud Agha entered and, seating himself near me, delivered a long speech to the effect that he was a servant of the Pasha, who was again the slave of the Sultan; and that servants were bound to obey the commands of their master, however disagreeable and unjust they might be. I saw at once to what this exordium was about to lead and was prepared for the announcement that he had received orders from Mosul to stop the excavations by threatening those who were inclined to work for me. On the following morning, therefore, I rode to the town and waited upon his Excellency. He pretended to be taken by surprise, disclaimed having given such orders and directed his secretary to write at once to the commander of the irregular troops, who was to give me every assistance rather than throw impediments in my way. He promised to let me have the letter in the afternoon before I returned to Selamiyah, but an officer came to me soon after and stated that as the Pasha was unwilling to detain me he would forward it in the night. I rode back to the village and acquainted Daoud Agha with the result of my visit. About midnight, however, he returned to me and declared that a horseman had just brought him more stringent orders than any he had yet received and that on no account was he to permit me to carry on the work.

Surprised at this inconsistency, I returned to Mosul early the next day and again called upon the Pasha. "It was with deep regret," said he, "I learned, after your departure yesterday, that the mound in which you are digging had been used as a burying ground by Moslems and was covered with their graves. Now you are aware that by the law it is forbidden to disturb a tomb, and the cadi and mufti [interpreter of Moslem law] have already made representations to me on the subject."

"In the first place," replied I, "being pretty well acquainted with the mound, I can state that no graves have been disturbed; in the second, gravestones would present no difficulty."

He returned, "No, I cannot allow you to proceed; you are my dearest and most intimate friend; if anything happens to you, what grief should I not suffer! Your life is more valuable than old stones; besides, the responsibility would fall upon my head." Finding that the Pasha had resolved to interrupt my proceedings, I pretended to acquiesce in his answer and requested that a cawass of his own might be sent with me to Nimroud, as I wished to draw the sculptures and copy the inscriptions which had already been uncovered. To this he consented and ordered an officer to accompany me.

On my return to Selamiyah there was little difficulty in inducing the Pasha's cawass to countenance the employment of a few workmen to *guard* the sculptures during the day; and as Daoud Agha considered that this functionary's presence relieved him from any further responsibility, he no longer interfered with any experiment I might think proper to make.

Wishing to ascertain the existence of the graves and also to draw one of the bas-reliefs which had been uncovered, though not to continue the excavations for a day or two, I rode to the ruins on the following morning, accompanied by the Hytas [irregular cavalry] and their chief, who were going their usual round in search of plundering Arabs. Daoud Agha confessed to me on our way that he had received orders to make graves on the mound and that his troops had been employed for two nights in bringing stones from distant villages for that purpose. "We have destroyed more real tombs of the true Believers," said he, "in making sham ones than you could have defiled between the Zab and Selamiyah. We have killed our horses and ourselves in carrying those accursed stones." A steady rain setting in, I left the horsemen and returned to the village.

I continued to employ a few men to open trenches by way of experiment, and was not long in discovering other sculptures. Near the western edge [of the Southwest Palace] we came upon the lower part of several gigantic figures, uninjured by fire. In the center of the mound [Center Palace] we uncovered part of a pair of gigantic winged bulls, the head and half the wings of which had been destroyed. On the back of the enormous slabs on which these animals had been carved were inscriptions in large and well-cut characters.

The experiment had been fairly tried. There was no longer any doubt of the existence not only of sculptures and inscriptions, but even of vast edifices in the interior of the mound of Nimroud, as all parts of it that had been examined furnished remains of buildings and carved slabs. I lost no time, therefore, in acquainting Sir Stratford Canning with my discovery, and urging the necessity of an order from the Porte [government at Constantinople] which would prevent any future interference on the part of the authorities or the inhabitants of the country.

It was now nearly Christmas, and as it was desirable to remove all the tombs, which had been made by the Pasha's orders, on the mound, I came to an understanding on the subject with Daoud Agha. I covered over the sculptures brought to light and withdrew altogether from Nimroud, leaving an agent at Selamiyah.

On entering Mosul on the morning of the eighteenth, I found the whole population in a ferment of joy. A Tartar had that morning brought from Constantinople the welcome news that the Porte, at length alive to the wretched condition of the province and to the misery of the inhabitants, had disgraced the governor and named Ismail Pasha, a young Major-General of the new school, to carry on affairs until Hafiz Pasha, the new governor, could reach the province.

During these events the state of the country rendered the continuation of my researches at Nimroud almost impossible. I determined, therefore, to proceed to Baghdad to make arrangements for the removal of the sculptures at a future period and to consult generally with Major Henry Creswicke Rawlinson, from whose experience and knowledge I could derive the most valuable assistance.

Chapter 3

■ On my return to Mosul in the beginning of January [1846], I found Ismail Pasha installed in the government. He received me with courtesy, offered no opposition to the continuation of my researches at Nimroud and directed the irregular troops stationed at Selamiyah to afford me every assistance and protection. The change since my departure had been as sudden as great. A few conciliatory acts on the part of the new Governor and a promise of diminution of taxes had so far reassured and gained the confidence of those who had fled to the mountains and the desert that the inhabitants of the villages were slowly returning to their homes; and even the Arab tribes which were formerly accustomed to pasture their flocks in the districts of Mosul were again pitching their tents on the banks of the Tigris.

I rode to Nimroud on the seventeenth of January, having first engaged a party of Nestorian Chaldeans to accompany me.

The change that had taken place in the face of the country during my absence was no less remarkable than that which I had found in the political state of the province. To me they were both equally agreeable and welcome. The rains, which had fallen almost incessantly from the day of my departure for Baghdad, had rapidly brought forward the vegetation of spring. The mound was no longer an arid and barren heap; its surface and its sides were covered with verdure. From the summit of the pyramid [ziggurat, or high temple tower] my eye ranged, on one side, over a broad level enclosed by the Tigris and the Zab; on the other, over a low, undulating country bounded by the snow-capped mountains of Kurdistan. But it was no longer the dreary waste I had left a month before. The landscape was clothed in green, the black tents of the Arabs checkered the plain of Nimroud and their numerous flocks pastured on the distant hills. Arabs had also returned to their villages and were now engaged in cultivating the soil. Even on the mound the plough opened its furrows and corn was sown over the palaces of the Assyrian kings.

Security had been restored, and Nimroud offered a more convenient and more agreeable residence than Selamiyah. Hiring three huts, which had been hastily built in the outskirts of the village, I removed to my new dwelling place. A few rude chairs, a table and a wooden bedstead formed the whole of my furniture. My cawass spread his carpet and hung his tobacco pouch in

The Metropolitan Museum of Art,
Gift of John D. Rockefeller, Jr., 1931 (detail)

Hormuzd Rassam, British Museum

the corner of a hovel which he had appropriated, and spent his days in peaceful contemplation. The servants constructed a rude kitchen, and the grooms shared the stalls with the horses. Mr. Hormuzd Rassam, the brother of the British Vice-Consul, came to reside with me and undertook the daily payment of the workmen and the domestic arrangements.

My agent, with the assistance of the chief of the Hytas, had punctually fulfilled the instructions he had received on my departure. Not only were the counterfeit graves carefully removed, but others, which possessed more claim to respect, had been rooted out. I entered into an elaborate argument with the Arabs on the subject of the latter and proved to them that, as the bodies were not turned toward Mecca, they could not be those of true Believers. I ordered the remains, however, to be carefully collected and to be reburied at the foot of the mound.

I had scarcely resumed my labors when I received information that the cadi of Mosul was endeavoring to stir up the people against me, chiefly on the plea that I was carrying away treasure. A representation was ultimately made to Ismail Pasha; and as he expressed a wish to see me, I rode to Mosul. He was not, he said, influenced by the cadi or the mufti, nor did he believe the absurd tales which they had spread abroad. I should shortly see how he intended to treat these troublesome fellows, but he thought it prudent at present to humor them and made it a personal request that I would, for the time, suspend the excavations. I consented with regret and once more returned to Nimroud without being able to gratify the ardent curiosity I felt to explore further.

It was near the middle of February before I thought it prudent to make some fresh experiments among the ruins. To avoid notice I only employed a few men and confined myself to the examination of such parts of the mound as appeared to contain buildings. A trench was first opened at right angles to the center of wall *k* [Southwest Palace], and we speedily found the wall *q*. All the slabs were sculptured, and uninjured by fire, but unfortunately had been half destroyed by long exposure to the atmosphere. It was evident from the costume, the ornaments and the nature of the relief that these sculptures did not belong either to the same building or to the same period as those previously discovered. I recognized in them the style of Khorsabad and in the inscriptions particular forms in the character which were used in the inscriptions of that monument. Still the slabs were not *in situ*; they had been brought from elsewhere, and I was even more perplexed than I had hitherto been.

These experiments were sufficient to prove that the building I was exploring had not been entirely destroyed by fire, but had been partly exposed to gradual decay. No sculptures had hitherto been discovered in a perfect state of preservation, and only one or two could bear removal. I determined, therefore, to abandon this corner and to resume excavations near the cham-

ber first opened (chamber A, Northwest Palace), where the slabs had in no way been injured. The workmen were directed to dig behind the small lions [chamber BB], which appeared to form an entrance and to be connected with other walls. After removing much earth, a few unsculptured slabs were discovered, fallen from their places and broken in many pieces. The sides of the room of which they had originally formed a part could not be traced.

As these ruins occurred on the edge of the mound, it was probable that they had been more exposed than the rest and consequently had sustained more injury than other parts of the building. As there was a ravine running far into the mound, apparently formed by the winter rains, I determined to open a trench in the center of it. In two days the workmen reached the top of a slab, which appeared to be both well preserved and to be still standing in its original position. On the south side [chamber B] I discovered, to my great satisfaction, two human figures, considerably above the natural size, sculptured in low relief, and still exhibiting all the freshness of a recent work. In a few hours the earth and rubbish had been completely removed from the face of the slab, no part of which had been injured. The ornaments delicately graven on the robes, the tassels and fringes, the bracelets and armlets, the elaborate curls of the hair and beard were all entire. The figures were back to back and furnished with wings. They appeared to represent divinities presiding over the seasons or over particular religious ceremonies. The one, whose face was turned to the East, carried a fallow deer on his right arm and in his left hand a branch bearing five flowers. Around his temples was a fillet, adorned in front with a rosette. The other held a square vessel, or basket, in the left hand and an object resembling a fir cone in the right. On his head he wore a rounded cap, at the base of which was a horn.

To the west of this slab, and fitting to it, was a cornerstone ornamented with flowers and scrollwork, tastefully arranged and resembling in detail those graven on the injured tablet near entrance *d* of the Southwest building. I recognized at once from whence many of the sculptures employed in the construction of that edifice had been brought, and it was evident that I had at length discovered the earliest palace of Nimroud [Northwest Palace].

The cornerstone led me to a figure of singular form [chamber B]. A human body, clothed in robes similar to those of the winged men on the previous slab, was surmounted by the head of an eagle or of a vulture. The curved beak, of considerable length, was half open and displayed a narrow

The Metropolitan Museum of Art,
Gift of John D. Rockefeller, Jr., 1931

pointed tongue which was still colored with red paint. On the shoulders fell the usual curled and bushy hair of the Assyrian images, and a comb of feathers rose on the top of the head. Two wings sprang from the back, and in either hand was the square vessel and fir cone.

On all these figures paint could be faintly distinguished, particularly on the hair, beard, eyes and sandals. The slabs on which they were sculptured had sustained no injury and could be without difficulty packed and moved to any distance. There could no longer be any doubt that they formed part of a chamber and that, to explore it completely, I had only to continue along the wall, now partly uncovered.

On the morning following these discoveries, I was returning to the mound when I saw two Arabs urging their mares to the top of their speed. On approaching me they stopped. "Hasten, O Bey," exclaimed one of them, "hasten to the diggers, for they have found Nimrod himself ["the Mighty Hunter" mentioned in Genesis 10:9]. Wallah, it is wonderful, but it is true! We have seen him with our eyes. There is no God but God," and both joining in this pious exclamation, they galloped off without further words in the direction of their tents.

On reaching the ruins I descended into the new trench [in the Northwest Palace] and found the workmen, who had already seen me as I approached, standing near a heap of baskets and cloaks. While Awad advanced and asked for a present to celebrate the occasion, the Arabs withdrew the screen they had hastily constructed and disclosed an enormous human head sculptured in full out of the alabaster of the country. They had uncovered the upper part of the figure, the remainder of which was still buried in the earth. I saw at once that the head must belong to a winged lion or bull similar to those of Khorsabad and Persepolis. It was in admirable preservation. The expression was calm yet majestic, and the outline of the features showed a freedom and knowledge of art scarcely to be looked for in the works of so remote a period. The cap had three horns and, unlike that of the human-headed bulls hitherto found in Assyria, was rounded and without ornament at the top.

I was not surprised that the Arabs had been amazed and terrified at this apparition. It required no stretch of imagination to conjure up the most strange fancies. This gigantic head, blanched with age, thus rising from the bowels of the earth might well have belonged to one of those fearful beings which are pictured in the traditions of the country as appearing to mortals,

Discovery of the gigantic stone head.
From Layard, Nineveh and Its Remains

slowly ascending from the regions below. One of the workmen, on catching the first glimpse of the monster, had thrown down his basket and run off toward Mosul as fast as his legs could carry him. I learned this with regret, as I anticipated the consequences.

I now ordered a trench to be dug due south from the head in the expectation of finding a corresponding figure, and before nightfall reached the object of my search about twelve feet distant. Engaging two or three men to sleep near the sculptures, I returned to the village and celebrated the day's discovery by a slaughter of sheep, of which all the Arabs near partook. As some wandering musicians chanced to be at Selamiyah, I sent for them, and dances were kept up during the greater part of the night. On the following morning Arabs from the other side of the Tigris and the inhabitants of the surrounding villages congregated on the mound. Even the women could not repress their curiosity and came in crowds, with their children, from afar. My cawass was stationed during the day in the trench into which I would not allow the multitude to descend.

As I had expected, the report of the discovery of the gigantic head, carried by the terrified Arab to Mosul, had thrown the town into commotion. He had scarcely checked his speed before reaching the bridge. Entering breathless into the bazaars, he announced to everyone he met that Nimrod had appeared. The news soon got to the ears of the cadi, who, anxious for a fresh opportunity to annoy me, called the mufti and the ulema [scholars] together, to consult upon this unexpected occurrence. Their deliberations ended in a procession to the Governor and a formal protest, on the part of the Moslems of the town, against proceedings so directly contrary to the laws of the Koran. The cadi had no distinct idea whether the bones of Nimrod the mighty hunter had been uncovered or only his image, nor did Ismail Pasha very clearly remember whether Nimrod was a true-believing prophet or an Infidel. I consequently received a somewhat unintelligible message from his Excellency to the effect that the remains should be treated with respect and be by no means further disturbed and that he wished the excavations to be stopped at once and desired to confer with me on the subject.

I called upon him accordingly and had some difficulty in making him understand the nature of my discovery. As he requested me to discontinue my operations until the sensation in the town had somewhat subsided, I returned to Nimroud and dismissed the workmen, retaining only two men to dig leisurely along the walls without giving cause for further interference.

I ascertained by the end of March the existence of a second pair of winged human-headed lions [chamber B of Northwest Palace]. They formed a northern entrance into the chamber.

I completely uncovered the lions on the southern portal and found them to be entire. They were about twelve feet in height and the same number in length. The body and limbs were admirably portrayed: the muscles and bones, although strongly developed to display the strength of the animal, showed at the same time a correct knowledge of its anatomy and form. Expanded wings sprung from the shoulder and spread over the back; a knotted girdle, ending in tassels, encircled the loins. These sculptures were partly in full and partly in relief. The head and forepart, facing the chamber, were in full, but only one side of the rest of the slab was sculptured, the back being placed against the wall of sun-dried bricks. That the spectator might have both a perfect front and side view of the figures, they were furnished with five legs; two were carved on the end of the slab to face the chamber, and three on the side. The relief of the body and three limbs was high and bold, and the slab was covered, in all parts not occupied by the image, with inscriptions in the cuneiform character. These magnificent specimens of Assyrian art were in perfect preservation; the most minute lines in the details of the wings and in the ornaments had been retained with their original freshness. Not a character was wanting in the inscriptions.

I used to contemplate for hours these mysterious emblems and muse over their intent and history. What more sublime images could have been borrowed from nature by men who sought to embody their conception of the wisdom, power and ubiquity of a Supreme Being? They could find no better type of intellect and knowledge than the head of the man; of strength, than the body of the lion; of rapidity of motion, than the wings of the bird. For twenty-five centuries they had been hidden from the eye of man and they now stood forth once more in their ancient majesty. But how changed was the scene around them! The wealth of temples and the riches of great cities had been succeeded by ruins and shapeless heaps of earth. Above the spacious hall in which they stood, the plough had passed and the corn now waved.

The operations at Nimroud having been completely suspended until I could receive assistance from Constantinople, I thought the time opportune to visit the Sheikh of the great Arab tribe which occupies nearly the whole of Mesopotamia.

Chapter 4

■ During my absence little progress had been made, as only two men had been employed in removing the rubbish from the upper part of the chamber [B in the Northwest Palace] to which the great human-headed lions formed an entrance. The lions to the east of them (entrance *d*) had, however, been completely uncovered; that to the right had fallen from its place and was sustained by the opposite sculpture. Between them was a large pavement slab covered with cuneiform characters.

It was remarkable that while such parts of the chamber B as had been uncovered were paved with kiln-burnt bricks, and the entrance *d* with a large slab of alabaster, between the two great lions there was only a flooring of common sun-dried brick.

As several of the principal Christian families of Mosul were anxious to see the sculpture, whose fame had spread over the town and provinces, I was desirous of gratifying their curiosity before the heat of the summer rendered the plain of Nimroud almost uninhabitable. An opportunity at the same time presented itself of securing the goodwill of the Arab tribes encamping near the ruins, by preparing an entertainment which might gratify all parties.

White pavilions, borrowed from the Pasha, had been pitched near the river, on a broad lawn carpeted with flowers. These were for the ladies, and for the reception of the Sheikh. Black tents were provided for some of the guests, for the attendants and for the kitchen. An open space was left in the center of the group of tents for dancing and for various exhibitions provided for the entertainment of the company.

Fourteen sheep had been roasted and boiled to feast the crowd that had assembled. They were placed on large wooden platters, which, after the men had satisfied themselves, were passed on to the women. The dinner having been devoured to the last fragment, dancing succeeded. A party of Kurdish jesters from the mountains entertained the Arabs with performances and imitations. They were received with shouts of laughter. The dances were kept up by the light of the moon the greater part of the night. The festivities lasted three days, and made the impression I had anticipated. They earned me a great reputation and no small respect.

Hafiz Pasha, who had been appointed to succeed the last governor, having received a more lucrative post, the province was sold to Tahyar Pasha. He

The Cleveland Museum of Art (detail)

made his public entry into Mosul early in May, and I rode out to meet him. He was followed by a large body of troops and by the cadi, mufti, ulema and principal inhabitants of the town, who had been waiting for him at some distance from the gates to show their respect. Good reports of his benevolence and justice had preceded him. He was a venerable old man, bland and polished in his manners, courteous to Europeans and well informed on subjects connected with the literature and history of his country. He was a perfect specimen of the Turkish gentleman of the old school, of whom few are now left in Turkey. I had been furnished with serviceable letters of introduction to him; he received me with every mark of attention and at once permitted me to continue the excavations. As a matter of form, he named a cawass to superintend the work on his part. I willingly concurred in this arrangement, as it saved me from any further inconvenience on the score of treasure, for which, it was still believed, I was successfully searching.

The heats of summer had now commenced and it was no longer possible to live under a white tent. The huts were equally uninhabitable and swarmed with vermin. In this dilemma I ordered a recess to be cut into the bank of the river where it rose perpendicularly from the water's edge. By screening the front with reeds and boughs of trees and covering the whole with similar materials, a small room was formed. I was much troubled, however, with scorpions and other reptiles which issued from the earth forming the walls of my apartment; and later in the summer by the gnats and sandflies which hovered on a calm night over the river.

The change to summer had been as rapid as that which ushered in the spring. The verdure of the plain had perished almost in a day. Hot winds coming from the desert had burned up and carried away the shrubs; flights of locusts, darkening the air, had destroyed the few patches of cultivation and had completed the havoc commenced by the heat of the sun. When on the mound, my only secure place of refuge was beneath the fallen lion, where I could defy the fury of the whirlwind; the Arabs ceased from their work and crouched in the trenches, almost suffocated and blinded by the dense cloud of fine dust and sand.

Although the number of my workmen was small, the excavations were carried on as actively as possible. The two human-headed lions, forming the entrance *d* [chamber B, Northwest Palace] led into another chamber, or to sculptured walls which may have formed an outward facing to the building. The slabs to the right and left, on issuing from this portal, had fallen from their original position and all of them except one were broken. I had some difficulty in raising the pieces from the ground. As the face of the slabs was downward, the sculpture had been well preserved.

On two of the slabs the king was represented holding a bow in one hand and two arrows in the other. He was followed by his attendant eunuch, who carried a mace, a second bow and a quiver for his use.

The first slab on the other side of the entrance contained a vizier and his attendant. The succeeding slabs were occupied by figures, differing altogether in costume from those previously discovered and apparently representing people of another race; some carrying presents or offerings consisting of armlets, bracelets and earrings on trays; others elevating their clinched hands, either in token of submission or in the attitude still peculiar to Easterns when they dance. One figure was accompanied by two monkeys held by ropes, the one raising itself on its hind legs in front, the other sitting on the shoulders of the man and supporting itself by placing its forepaws on his head. The dresses of all these figures are singular. They have high boots turned up at the toes. The figure with the monkeys wears a tunic descending to the calf of the leg. His hair is simply fastened by a fillet. These peculiarities of dress suggest that the persons represented were captives from some distant country bringing tribute to the conquerers.

In chamber B the wall was continued to the south, or to the left facing the great lion, by an eagle-headed figure.

The first sculpture, still standing in its original position, which was uncovered after following this wall was a winged human-headed bull of yellow limestone. On the previous day its detached head had been found. The bull had fallen against the opposite sculpture and had been broken by the fall into several pieces.

Beyond the winged bull the slabs were still entire and occupied their original positions. On the first was sculptured a winged human figure carrying a branch with five flowers in the raised right hand and the usual square vessel in the left.

In the upper compartment of the next slab was the siege of the city with the battering ram and movable tower. The lower part of the two slabs was occupied by one subject, a king receiving prisoners brought before him by his vizier. The prisoners were on the adjoining slab. Above their heads are vases and various objects among which appear to be shawls and elephants' tusks, probably representing the spoil carried away from the conquered nation.

British Museum

Upon the third slab were, in the upper compartment, the king hunting, and in the lower, the king standing over the lion; and on the fourth the bull hunt and the king standing over the prostrate bull.

On the flooring below the sculptures were discovered considerable remains of painted plaster still adhering to the sun-dried bricks, which had fallen in masses from the upper part of the wall. The colors, particularly the blues and reds, were as brilliant and vivid when the earth was removed from them as they could have been when first used. On exposure to the air they faded rapidly. The designs were elegant and elaborate. It was found almost impossible to preserve any portion of these ornaments, the earth crumbling to pieces when an attempt was made to raise it.

About this time I received the vizierial letter procured by Sir Stratford Canning authorizing the continuation of the excavations and the removal of such objects as might be discovered. I was sleeping in the tent of Sheikh Abd-ur-rahman, who had invited me to hunt gazelles with him before dawn on the following morning, when an Arab awoke me. He was the bearer of letters from Mosul, and I read by the light of a small camel-dung fire the document which secured to the British nation the records of Nineveh and a collection of the earliest monuments of Assyrian art.

One of the difficulties, and not one of the least which had to be encountered, was now completely removed. Still, however, pecuniary resources were wanting, and in the absence of the necessary means, extensive excavations could not be carried on. I hastened nevertheless to communicate the letter of the Grand Vizier to the Pasha and to make arrangements for pursuing the researches as effectually as possible.

Not having yet examined the great mound of Kouyunjik, which has generally been believed by travelers to mark the true site of Nineveh, I determined to open trenches in it. I had not previously done so as the vicinity of the ruins to Mosul would have enabled the inhabitants of the town to watch my movements and to cause me continual interruptions before the sanction of the authorities could be obtained to my proceedings. A small party of workmen having been organized, excavations were commenced on the southern face, where the mound was highest, as sculptures, if any still existed, would probably be found in the best state of preservation under the largest accumulation of rubbish.

The only opposition I received was from the French Consul, who claimed the ruins as French property. The claim not being recognized, he also dug into the mound, but in another direction. We both continued our researches for about a month without much success.

On my return to Nimroud, about thirty men, chiefly Arabs, were employed to carry on the excavations. Being anxious to learn as soon as possible the extent of the building [chamber B, Northwest Palace] and the nature of the sculptures it contained, I merely dug down to the top of the slabs and ascertained the character of the sculpture upon them, reserving a completer examination for a more favorable opportunity. I was thus able to form an opinion as to the number of bas-reliefs that could be removed and to preserve those partially uncovered from injury by heaping the rubbish again over them.

United to the last of the four slabs with small bas-reliefs, beyond the bulls of yellow limestone, was an ornamented cornerstone marking the end of hall B. Adjoining the cornerstone was a winged figure.

The Arabs marveled at these strange figures. As each head was uncovered they showed their amazement by extravagant gestures or exclamations of surprise. If it was a bearded man, they concluded at once that it was an idol or a genie and cursed or spat upon it. If a eunuch, they declared that it was the likeness of a beautiful female and kissed or patted the cheek. They soon felt as much interest as I did in the objects discovered and worked with renewed ardor when their curiosity was excited by the appearance of a fresh sculpture. On such occasions they would strip themselves almost naked, throw the handkerchief from their heads and, letting their matted hair stream in the wind, rush like madmen into the trenches and carry off the baskets of earth, shouting at the same time the war cry of the tribe.

On other uncovered slabs of chamber G [Northwest Palace] were groups composed of the king, raising the cup and attended by two eunuchs, or holding a bow in one hand and two arrows in the other, preceded and followed by a winged human figure. These groups were alternate and were all equally remarkable for the richness and elegance of the embroidery and ornaments. They furnished me not only with a collection of beautiful designs but also with many new and highly interesting symbolical and mythic signs and figures.

I did not, for the time, follow the eastern wall of this chamber, but turned into an entrance which led me into another chamber (H). This entrance was formed by two winged figures, and two plain slabs.

I quitted this chamber after uncovering the upper part of four or five bas-reliefs, and returning to an entrance to chamber G, traced to the south of it two slabs upon which were groups similar to those on the opposite wall, except that the right hand of the king rested on the hilt of his sword and not on the bow. One was an eagle-headed figure, and beyond it was discovered another pair of human-headed lions. The slabs on which they were carved were slightly cracked, but otherwise they appeared to have issued but the day before from the hand of the sculptor. The accumulation of earth and rubbish above this part of the ruins was very considerable and it is not improbable that it was owing to this fact that the sculptures had been so completely guarded from injury.

Without proper materials it was impossible to move either the gigantic lions or even the large sculptures. The few ropes to be obtained in the country were so ill made that they could not support any considerable weight. I determined, therefore, to displace the slabs divided into two compartments, then to saw off the sculptures and to reduce them as much as possible by cutting from the back. The inscriptions being mere repetitions I did not consider it necessary to preserve them, as they added to the weight. With the help of levers of wood, and by digging away the wall of sun-dried bricks from behind the slabs, I was enabled to turn them into the center of the trench where they were sawn by marble-cutters from Mosul. When the bas-reliefs were thus prepared, there was no difficulty in dragging them out of the trenches.

The Metropolitan Museum of Art, Gift of John D. Rockefeller, Jr., 1932

After having been removed from the trenches, the sculptures were packed in felts and matting and screwed down in roughly made cases. They were transported from the mound to the river upon rude buffalo carts belonging to the Pasha and then placed upon a raft formed of inflated skins and beams of poplar wood. They floated down the Tigris as far as Baghdad, were there placed on board boats of the country and reached Busrah in the month of August.

The heat had now become so intense that my health began to suffer from continual exposure to the sun and from the labor of superintending the excavations, drawing the sculptures and copying the inscriptions. In the trenches, where I daily passed many hours, the thermometer generally ranged from 112° to 115° in the shade, and on one or two occasions even reached 117°. The hot winds swept over the desert; they were as blasts from a furnace during the day and even at night they drove away sleep.

The state of my health compelled me to renounce for a time my labors at Nimroud. As I required a cooler climate, I determined to visit the Tiyari Mountains, inhabited by the Chaldean Christians, and to return to Mosul in September, when the violence of the heat had abated.

Chapter 5

■ On my return to Mosul I received letters from England informing me that Sir Stratford Canning had presented the sculptures discovered in Assyria to the British nation and that the British Museum had received a grant of funds for the continuation of the researches commenced at Nimroud, and elsewhere. The grant was small and scarcely adequate. I determined, however, to economize as far as it was in my power that the British nation might possess as extensive and complete a collection of Assyrian antiquities as, considering the smallness of the means, it was possible to collect.

I had neither knowledge nor experience as a draftsman, and this I felt to be a great drawback, indeed a disqualification, which I could scarcely hope to overcome. Many of the sculptures and monuments discovered were in too dilapidated a condition to be removed, and others threatened to fall to pieces as soon as uncovered. It was only by drawings that the record of them could be preserved. There was no inclination to send an artist to assist me, and I made up my mind to do the best I could; to copy as carefully and accurately as possible that which I saw before me. I had therefore to superintend the excavations, to draw all the bas-reliefs discovered, to copy and compare the innumerable inscriptions, to take casts of them (casts of the inscriptions and some of the sculptures were taken with brown paper simply damped and impressed on the slab with a hard brush), and to preside over the moving and packing of the sculptures. As there was no one whom I could trust to overlook the diggers, I was obliged to be continually present and frequently to remove the earth myself from the face of the slabs as, through the carelessness and inexperience of the workmen, they were exposed to injury from the blows of the picks. I felt that I was far from qualified to undertake these multifarious occupations. I knew, however, that if persons equal to the task and sufficiently well acquainted with the various languages of the country to carry on the necessary communications with the authorities and to hold the requisite intercourse with the inhabitants—Arabs, Kurds, Turks and Chaldeans—were sent out expressly from England, the whole sum granted would be expended before the excavations could be commenced. The researches would probably be then less extensive, and their results less complete than they would be if, however unqualified, I at once undertook their superintendence. I determined, therefore, to devote the whole of my time to the undertaking and to make every sacrifice to ensure its success.

The Metropolitan Museum of Art,
Gift of John D. Rockefeller, Jr., 1931 (detail)

It was, in the first place, necessary to organize a band of workmen best fit to carry on the work. I had no difficulty in finding workmen. There was this advantage in employing wandering Arabs—they brought their tents and families with them and, encamping round the ruins and the village, formed a very efficient guard against their brethren of the desert who looked to plunder. To increase my numbers, I chose only one man from each family; and, as his male relations accompanied him, I had their services for the protection of my sculptures. The Arabs were selected to remove the earth—they were unable to dig; this part of the labor required stronger and more active men, and I chose for it about fifty Nestorian Chaldeans, many of whom, having already been employed, had acquired some experience in excavating. I engaged at the same time a Jacobite, or Syrian Christian, who was a skillful marble cutter and a very intelligent man. I had made also a valuable addition to my establishment. His name was Mohammed Agha, but he was generally called, from his rank, the "Bairakdar" (a standard-bearer of the irregular troops). He served me faithfully and honestly and was of great use during the excavations. Awad still continued in my employ; my cawass, Ibrahim Agha, returned with me and I hired a carpenter and two or three men as superintendents.

I was again among the ruins by the end of October. The winter season was fast approaching, and it was necessary to build a proper house for the shelter of myself and servants. I marked out a plan on the ground on the outside of the village of Nimroud, and in a few days the habitations were complete. My workmen formed the walls of mud bricks dried in the sun, covering the rooms with beams and branches of trees. A thick coat of mud was laid over the whole to exclude the rain. I built two rooms for myself, divided by an open apartment, the whole being surrounded by a wall. In a second courtyard were huts for my cawass, for Arab guests and for my servants, and stables for my horses. Ibrahim Agha displayed his ingenuity by making loopholes of a most warlike appearance in the walls. I immediately ordered these to be filled up to avoid any suspicion of being the constructor of forts and castles. We did not neglect precautions, however, in case of an attack from the Bedouins, of whom Ibrahim Agha was in constant dread. Unfortunately, the only showers of rain that I saw during the remainder of my residence in Assyria fell before my walls were covered and so saturated the bricks that they did not become dry before the following spring. The consequence was that the only verdure on which my eyes were permitted to feast

Layard sketching in the Hall of Archives, Kuyunjik. From Layard, Nineveh and Babylon, *Photograph, British Museum*

before my return to Europe was furnished by my own property—the walls in the interior of the rooms being continually clothed with a crop of grass.

On the mound itself, and immediately above the great winged lions first discovered, I built a house for my Nestorian workmen and their families and a hut to which I could at once remove for safety any small objects discovered among the ruins. I divided my Arabs into three parties. About forty tents were pitched on different parts of the mound at the entrances to the principal trenches. Forty more were placed round my dwelling and the rest on the bank of the river, where the sculptures were deposited previous to their embarkation on the rafts. The men were all armed. I thus provided for the defense of all my establishment.

Mr. Hormuzd Rassam lived with me, and to him I confided the payment of the wages and all the accounts.

I divided my workmen into parties. In each set were generally eight or ten Arabs who carried away the earth in baskets; and two or four Nestorian diggers, according to the nature of the soil and rubbish which had to be removed. They were overlooked by a superintendent whose duty it was to keep them to their work and to give me notice when the diggers approached any slab or exposed any small object to view, that I might myself assist in the uncovering or removal. I scattered a few Arabs of a hostile tribe among the rest, and by that means I was always made acquainted with what was going on, could easily learn if there were plots brewing and could detect those who might attempt to appropriate any relics discovered during the excavations.

The smallness of the sum placed at my disposal compelled me to follow the same plan in the excavations that I had hitherto adopted—namely, to dig trenches along the sides of the chambers and to expose the whole of the slabs without removing the earth from the center. Thus, few of the chambers were fully explored, and many small objects of great interest may have been left undiscovered. As I was directed to bury the building with earth after I had explored it, to avoid unnecessary expense I filled up the chambers with the rubbish taken from those subsequently uncovered, having first examined the walls, copied the inscriptions and drawn the sculptures.

The excavations were recommenced on a large scale by the first of November. My working parties were distributed over the mound. I also opened trenches in parts of the ruins hitherto unexamined.

It will be remembered that in chamber B [of the Northwest Palace] the slabs had fallen with their faces to the ground. I was, in the first place, anxious to raise these bas-reliefs and to pack them for removal. To accomplish this I had to remove a large accumulation of earth and rubbish—to empty, indeed, almost the whole chamber, for the fallen slabs extended almost halfway across it. Some of the sculptures were found to be in admirable preservation although the slabs were broken by the fall.

The bas-reliefs were of the highest interest. They represented the wars of the king and the conquest of some strange people. The two upper bas-reliefs formed one subject—the king followed by warriors, in battle with his enemies under the walls of a hostile castle. He stands, gorgeously attired, in a chariot drawn as usual by three horses richly caparisoned. He is discharging an arrow.

Behind the king are three chariots; the first, drawn by three horses—one of which is rearing and another falling—is occupied by a warrior already pierced by an arrow and apparently demanding quarter of the pursuers. In the two other chariots are two warriors, one discharging an arrow, the other guiding the horses which are at full speed. In each chariot is a standard.

On the two next slabs was the return after victory. In front of the procession are several warriors carrying heads and throwing them at the feet of the conquerors. Two musicians are playing with a plectrum on stringed instruments or harps. They are followed by the warriors, who were represented in battle in the previous bas-relief, now unarmed and holding their standards before them; above them flies an eagle with a human head in his talons. Behind them is the king carrying in one hand his bow and in the other two arrows—the position in which he is so frequently represented on Assyrian monuments, and probably denoting triumph over his enemies.

After the procession we have the castle and pavilion of the conquering king. The ground plan of the former is represented by a circle, divided into four equal compartments and surrounded by towers and battlements. In each compartment there are figures apparently engaged in various culinary occupations and preparing the feast.

The four following bas-reliefs represent a battle in which the king, the two warriors with their standards and a eunuch are represented in chariots; and four warriors, among whom is also a eunuch, are on horses. The enemy are on foot, some wounded, some dead; others discharging their arrows against the pursuers. Eagles fly above the victors, and one is already feeding on a dead body. The winged divinity in the circle is again seen above the king. These bas-reliefs are executed with great spirit, particularly that containing the horsemen.

On the lower series of bas-reliefs are represented three subjects—the siege of a castle, the king receiving his prisoners, and the king with his army crossing a river. The first occupies three slabs. The greater part of the castle is in the center bas-relief. It has three towers, apparently several walls, one behind the other. They are all surmounted by angular battlements. The besiegers have brought a battering ram, attached to a movable tower, up to the wall, from which many stones have already been dislodged. One of the besieged has succeeded in catching the ram by a chain and is endeavoring to raise or move it from its place, while two warriors of the assailing party are holding it down by hooks to which they are hanging. Another is throwing fire (traces of the red paint being still retained in the sculpture) from above upon the engine, which the besiegers endeavor to quench by pouring water upon it from two spouts in the movable tower. Two figures in full armor are undermining the walls with instruments like blunt spears, while two others appear to have found a secret passage into the castle. Three of the besieged are falling from the walls, and upon one of the towers are two women tearing their hair and extending their hands as if in the act of asking for mercy. The enemy are already mounting to the assault, and scaling ladders have been placed against the walls. The king, discharging an arrow and defended by a warrior in complete armor who holds a shield before him, stands on one side of the castle. He is attended by two eunuchs.

On the other side of the castle are two kneeling figures, one discharging an arrow, the other holding a shield for his companion's defense. Behind them is the vizier, also discharging an arrow and protected by the shield of a second warrior. He is followed by three more warriors.

Upon the lower part of the three next slabs is the king receiving captives. Behind the chariot of the king are two other chariots each containing a charioteer alone: they are passing under the walls of a castle.

The three remaining bas-reliefs are highly interesting and curious. The first represents a boat containing a chariot in which is the king. In one hand he holds two arrows, in the other a bow. A eunuch, standing in front of the chariot, is talking with the king and is pointing with his right hand to some object in the distance, perhaps the stronghold of the enemy. Behind the chariot is a second eunuch holding a bow and a mace. The boat is towed by two naked men who are walking on dry land, and four men row the vessel with oars. One oar, with a broad flat extremity, is passed through a rope hung round a thick wooden pin at the stern, and serves both to guide and impel the boat. A charioteer standing in the vessel holds by the halters four horses who are swimming over the stream. A naked figure is supporting himself upon an inflated skin. The water is shown by undulating lines covering the face of the slab. On the next slab are two smaller boats: in the first appears to be the couch of the king and a jar or large vessel, in the other an empty chariot; they are each impelled by two men, seated face to face at their oars. Five other figures, two leading horses by their halters, are swimming on skins. Two fish are represented in the water. On the third slab is the embarkation—men are placing two chariots in a boat which is about to leave the shores, two warriors are already swimming over, and two others are filling and tying up their skins on the bank. Behind them, on dry land, are three figures erect, probably officers superintending the proceedings, one of whom, a eunuch, holds a whip in his right hand which may have been used to keep the soldiers to their duty.

Chamber I [Northwest Palace] had only been partly uncovered; the slabs were still half buried, and the pavement was covered to the height of six feet. A party of Arabs was employed in removing the remaining earth. As we approached the floor, a large quantity of iron was found among the rubbish, and I soon recognized in it the scales of the armor represented on the sculptures. Each scale was separate and was of iron, from two to three

British Museum. Photograph by Alfred Tamarin

inches in length, rounded at one end and squared at the other, with a raised or embossed line in the center.

The iron was covered with rust and in so decomposed a state that I had much difficulty in cleansing it from the soil. Two or three baskets were filled with these relics.

While I was collecting and examining these curious relics, a workman, digging away the earth from a corner of the chamber, came upon a perfect vase, but unfortunately struck it with his pick and broke away the upper part of it. I took the instrument and, working cautiously myself, was rewarded by the discovery of two small vases, one in alabaster, the other in glass (both in the most perfect preservation), of elegant shape and admirable workmanship. Each bore the name and title of the Khorsabad king, written in two different ways, as in the inscriptions of Khorsabad.

It was in the center of the mound that one of the most remarkable discoveries awaited me. I have already mentioned the pair of gigantic winged bulls, first found there. They appeared to form an entrance and to be only part of a large building [Center Palace]. The inscriptions upon them contained a name, differing from that of the king, who had built the Northwest Palace. On digging farther I found a brick on which was a genealogy, the new name occurring first and as that of the son of the founder of the earlier edifice. This was to a certain extent a clue to the comparative date of the newly discovered building.

I now sought for the wall which must have been connected with the bulls. I dug round these sculptures and found no other traces of building except a few squared stones, fallen from their original places. As the backs of the bulls were completely covered with inscriptions in large and well-formed cuneiform characters, I was led to believe that they might originally have stood alone. Still there must have been other slabs near them. I directed a deep trench to be carried, at right angles, behind the northern bull. After digging about ten feet, the workmen found a slab lying flat on the brick pavement and having a gigantic winged figure sculptured in relief upon it. It resembled some already described and carried the fir cone and the square basket or utensil, but there was no inscription across it. Beyond was a similar figure still more gigantic in its proportions, being about fourteen feet in height. The trench was carried on in the same direction for several days but nothing more appeared.

I had business in Mosul and was giving directions to the workmen to guide them during my absence. Standing on the edge of the hitherto unprofitable trench, I doubted whether I should carry it any farther, but made up my mind not to abandon it until my return, which would be on the following day. I mounted my horse but had scarcely left the mound when a corner of black marble was uncovered, lying on the very edge of the trench. This attracted the notice of the superintendent of the party digging, who ordered the place to be further examined. The corner was part of an obelisk, about seven feet high, lying on its side ten feet below the surface.

An Arab was sent after me without delay to announce the discovery, and on my return I found the obelisk completely exposed to view. I descended eagerly into the trench and was immediately struck by the singular appearance and evident antiquity of the remarkable monument before me. We raised it from its recumbent position and with the aid of ropes speedily dragged it out of the ruins. Although its shape was that of an obelisk, yet it was flat at the top and cut into three gradins [low steps]. It was sculptured on the four sides; there were in all twenty small bas-reliefs and above, below and between them was carved an inscription 210 lines in length. The whole was in the best preservation; scarcely a character of the inscription was wanting, and the figures were as sharp and well defined as if they had been carved but a few days before. The king is twice represented, followed by his attendants; a prisoner is at his feet, and his vizier and eunuchs are introducing men leading various animals and carrying vases and other objects of tribute on their shoulders or in their hands. The animals are the elephant, the rhinoceros, the camel, the wild bull, the lion, a stag and various kinds of monkeys. Among the objects carried by the tribute-bearers may perhaps be distinguished the tusks of the elephant, shawls and some bundles of precious wood. From the nature, therefore, of the bas-reliefs, it is natural to conjecture that the monument was erected to commemorate the conquest of India or of some country far to the east of Assyria and on the confines of the Indian peninsula. The name of the king, whose deeds it appears to record, is the same as that on the center bulls, and it is introduced by a genealogical list containing many other royal names.

I lost no time in copying the inscriptions and drawing the bas-reliefs upon this precious relic. It was then carefully packed, to be transported at once to Baghdad. A party of trustworthy Arabs were chosen to sleep near it at night, and I took every precaution against the superstitions and prejudices of the natives of the country and the jealousy of rival antiquaries.

Discoveries of scarcely less interest and importance were made almost at the same time. The workmen were exploring walls [in the Southwest Palace]. On reaching the end of them, they discovered a pair of winged lions of which the upper part, including the head, was almost entirely destroyed. They differed in many respects from those forming the entrances of the Northwest Palace. They had but four legs; they were carved out of a coarse limestone and not out of alabaster.

The plan and nature of the edifice in the southwest corner [Southwest Palace] was still a mystery to me. All the slabs hitherto uncovered had evidently been brought from another building, chiefly from that in the northwest part of the mound [Northwest Palace]. I had not yet had sufficient experience in the Assyrian cuneiform character to draw any inference from the inscriptions occurring on the bricks found among the ruins in this part of the mound, so as to connect the name of the king upon them with that of the founder of any known building.

There were no inscriptions between the legs and behind the bodies of the lions just described, as in other buildings at Nimroud and Khorsabad. I had not yet found any sculptures unaccompanied by the name and genealogy of the founders of the edifice in which they had been placed. When no inscription was on the face, it was invariably to be found on the back of the slab. I determined, therefore, to dig behind the lions. I was not disappointed in my search; a few lines in the cuneiform character were discovered, and I recognized at once the names of three kings.

While excavations were thus successfully carried on in the center [of the mound], and among the ruins of the two palaces first opened [Northwest and Southwest], discoveries of a different nature were made in the southeast corner, which was much higher than any other part of the mound.

One morning, the superintendent of the workmen informed me that a slab had been uncovered bearing an inscription. I hastened to the spot and saw the stone he had described lying at the bottom of the trench. Upon it was the name of the same king, written as on the center bulls. The slab having been partly destroyed, the inscription was imperfect. I ordered it to be raised with the intention of copying it. This was quickly effected with the aid of an iron crow when, to my surprise, I found that it had been used as the lid of an earthen sarcophagus which, with its contents, was still entire beneath. The sarcophagus was about five feet in length and very narrow. The skeleton was well preserved but fell to pieces almost immediately on exposure to the

British Museum

air; by its side were two jars in baked clay of a red color and a small alabaster bottle, all precisely resembling in shape similar vessels discovered in Egyptian tombs. There was no other clue to the date or origin of the sepulcher.

The sarcophagus was too small to contain a man of ordinary size when stretched at full length, and it was evident from the position of the skeleton that the body had been doubled up when forced in. A second earthen case was soon found, differing in form from the first. It resembled a dish cover and was scarcely four feet long. In it were other vessels of baked clay. Its lid was a slab taken from some building, like the lid of the sarcophagus first discovered. Although the skulls were entire when first exposed to view they crumbled into dust as soon as touched, and I was unable to preserve either of them.

These six weeks were among the most prosperous and fruitful during my researches in Assyria. Every day produced some new discovery. My Arabs entered with alacrity into the work and felt almost as much interested in its results as I did myself. They were now well organized, and I had no difficulty in managing them. Even their private disputes and domestic quarrels were referred to me.

As for myself, I rose at daybreak and after a hasty breakfast rode to the mound. Until night I was engaged in drawing the sculptures, copying and molding the inscriptions and superintending the excavations and the removal and packing of the bas-reliefs. On my return to the village, I was occupied till past midnight in comparing the inscriptions with the paper impressions, in finishing drawings and in preparing for the work of the following day. Such was our manner of life during the excavations at Nimroud.

Early in December I had collected a sufficient number of bas-reliefs to load another raft, and I consequently rode into Mosul to make preparations for sending a second cargo to Baghdad. I soon procured all that was necessary for the purpose, and loading a small raft with spars and skins for the construction of a larger raft and with mats and felts for packing the sculptures, I returned to Nimroud.

The raftmen, having left Mosul late in the day, were afraid to cross the dam in the dark; they therefore tied the raft to the shore and went to sleep. They were attacked during the night and plundered. I appealed to the authorities, but in vain. The Arabs of the desert, they said, were beyond their reach. If this robbery passed unnoticed, the remainder of my property, and even my person, might run some risk. Besides, I did not relish the reflection that the mats and felts destined for my sculptures were now furnishing the tents of some Arab Sheikh.

Three or four days elapsed before I ascertained who were the robbers. They belonged to a small tribe encamping at some distance from Nimroud —notorious in the country for their thieving propensities. Having ascertained the position of their tents, I started off one morning at dawn, accompanied by Ibrahim Agha, the Bairakdar, and another irregular horseman who was in my service. We reached the encampment after a long ride and found the number of the Arabs to be greater than I expected. The arrival of strangers drew together a crowd which gathered round the tent of the Sheikh where I seated myself. A slight bustle was apparent in the women's department. I soon perceived that attempts were being made to hide various ropes and felts, the ends of which, protruding from under the canvas, I had little difficulty in recognizing.

"Peace be with you!" said I, addressing the Sheikh, who showed by his countenance that he was not altogether ignorant of the object of my visit. "Your health and spirits are, please God, good. We have long been friends, although it has never yet been my good fortune to see you. I know the laws of friendship; that which is my property is your property, and the contrary. But there are a few things, such as mats, felts and ropes, which come from afar and are very necessary to me, while they can be of little use to you; otherwise God forbid that I should ask for them. You will greatly oblige me by giving these things to me."

"As I am your sacrifice, O Bey," answered he, "no such things as mats, felts or ropes were ever in my tents [I observed a new rope supporting the principal pole]. Search, and if such things be found we give them to you willingly."

"Wallah, the Sheikh has spoken the truth," exclaimed all the bystanders.

"That is exactly what I want to ascertain; and, as this is a matter of doubt, the Pasha must decide between us," replied I, making a sign to the Bairakdar who had been duly instructed how to act. In a moment he had handcuffed the Sheikh and, jumping on his horse, dragged the Arab at an uncomfortable pace out of the encampment. "Now, my sons," said I, mounting leisurely, "I have found a part of that which I wanted; you must search for the rest."

They looked at one another in amazement. Although the Arabs were well armed, they were too much surprised to make any attempt at resistance, or perhaps they feared too much for their Sheikh, still jolting away at an uneasy pace in the iron grasp of the Bairakdar, who had put his horse to a brisk trot and held his pistol cocked in one hand. The women, swarming out of the tents, now took part in the matter. Gathering round my horse, they kissed the tails of my coat and my shoes, making the most dolorous supplications. I was not to be moved, however, and extricating myself with difficulty from the crowd, I rejoined the Bairakdar, who was hurrying on his prisoner with evident good will.

The Sheikh had already made himself well known to the authorities by his dealings with the villages, and there was scarcely a man in the country who could not bring forward a specious claim against him—either for a donkey, a horse, a sheep or a copper kettle. He was consequently most averse to an interview with the Pasha and looked with evident horror on the prospect of a journey to Mosul. I added considerably to his alarm by dropping a few friendly hints on the advantage of the dreary subterranean lock-up house under the governor's palace, and of the pillory and sticks. By the time he reached Nimroud he was fully alive to his fate and deemed it prudent to make a full confession. He sent an Arab to his tents, and next morning an ass appeared in the courtyard bearing the missing property, with the addition of a lamb and a kid by way of a conciliatory offering. I dismissed the Sheikh with a lecture and had afterward no reason to complain of him or of his tribe—nor indeed of any tribes in the neighborhood, for the story got abroad and was invested with several horrible facts in addition, which could only be traced to the imagination of the Arabs but which served to produce the effect I desired—a proper respect for my property.

By the middle of December a second cargo of sculptures was ready to be sent to Baghdad. I was again obliged to have recourse to the buffalo carts of the Pasha, and as none of the bas-reliefs and objects to be moved were of great weight, these rotten and unwieldy vehicles could be patched up for the occasion. On Christmas day I had the satisfaction of seeing a raft bearing twenty-three cases, in one of which was the obelisk, floating down the river. I watched them until they were out of sight and then galloped into Mosul to celebrate the festivities of the season with the few Europeans whom duty or business had collected in this remote corner of the globe.

Chapter 6

■ One morning Ibrahim Agha came to me with his eyes full of tears and announced the death of Tahyar Pasha.

Essad Pasha was at length appointed to succeed Tahyar, and soon after reached his Pashalic [territory of the Pasha]. These changes did not affect my proceedings. Armed with my firman I was able to defy the machinations of the cadi and the ulema, who did not cease their endeavors to throw obstacles in my way.

After the celebration of Christmas I returned to Nimroud, and the excavations were again carried on with activity.

The Northwest Palace was naturally the most interesting portion of the ruins, and to it were principally directed my researches. I had satisfied myself beyond a doubt that it was the most ancient building yet explored in Assyria. Not having been exposed to a conflagration like other edifices, the sculptures, bas-reliefs and inscriptions which it contained were still admirably preserved.

When the excavations were resumed, eight chambers were discovered [in the Northwest Palace]. By the end of the month of April [1847] I had explored almost the whole building and had opened twenty-eight chambers cased with alabaster slabs. The principal part of the edifice seems to have been that to the north. Chambers B and G contained the most remarkable bas-reliefs; they represented the deeds of the king in war and in the chase, his triumphant return and the celebration of religious ceremonies. The best artists had evidently been employed upon them, and they excelled all those that had yet been discovered in the elegance and finish of the ornaments and in the knowledge of art displayed in the grouping of the figures.

A certain symmetry was observed in the plan of the building, particularly in the arrangement of the chambers to the east.

In chamber H all the groups were similar—representing the king holding a cup in one hand and his bow in the other, attended by two winged figures with garlands round their heads. The sculptures in chamber G were chiefly remarkable for the variety and elegance of the ornaments on the robes of the king and his attendants. Among them were: the king slaying the lion and hunting the bull, winged figures before the sacred tree, religious emblems, various animals and elaborate scrollwork—all furnishing not only beautiful designs but important illustrations of the religion of the Assyrians.

An entrance to chamber G was formed by two gigantic eagle-headed

The Metropolitan Museum of Art,
Gift of John D. Rockefeller, Jr., 1931 (detail)

winged figures of considerable beauty and finish. In the chamber beyond were repeated the winged divinities or priests with the emblematic tree.

On each of the slabs forming a narrow passage, leading from the chamber P, were two winged figures back to back. They were well designed and carefully executed. Beyond them, on slabs in chamber S, was the king between two eunuchs. The figure of the king was one of the most carefully sculptured and best preserved in the palace. He is represented with one hand on the hilt of his sword, the other being supported by a long wand or staff.

On the remaining slab of chamber S the winged figures were repeated. Some carried flowers of various shapes while others had the usual fir cone and square basket or utensil.

Three sides alone of the great hall Y were found entire. From its size it is probable that it was not roofed in but was an open court. It appears to have been nearly square, but the western wall has been completely destroyed, the slabs having perhaps been carried away to be used in the construction of the Southwest Palace. Three entrances are still standing, one formed by a pair of winged lions, and two by winged bulls.

Behind the great hall Y, to the south, were a cluster of chambers leading one into another. Their proportions were small. V and W did not contain sculptures.

The chamber V is remarkable for the discovery, near the entrance *a*, of a number of ivory ornaments of considerable beauty and interest. These ivories, when uncovered, adhered so firmly to the soil and were in so forward a state of decomposition that I had the greatest difficulty in extracting them even in fragments. I spent hours lying on the ground, separating them with a penknife from the rubbish by which they were surrounded. The ivory separated itself in flakes. Even the falling away of the earth was sufficient to reduce it almost to powder. With all the care that I could devote to the collection of the fragments, many were lost, or remained unperceived, in the immense heap of rubbish under which they were buried. Since they have been in England, they have been admirably restored and cleaned. The gelatinous matter by which the particles forming the ivory are kept together had, from the decay of centuries, been completely exhausted. By an ingenious process it has been restored, and the ornaments, which on their discovery almost fell to pieces upon mere exposure to the air, have regained the appearance and consistency of recent ivory.

The most interesting are the remains of two small tablets, one nearly

The Metropolitan Museum of Art,
Gift of John D. Rockefeller, Jr., 1931

British Museum

entire, the other much injured. Upon them are represented two sitting figures holding in one hand the Egyptian scepter or symbol of power. Between them is a cartouche containing a name or words in hieroglyphs and surmounted by a feather or plume. The chairs on which the figures are seated, the robes of the figures themselves, the hieroglyphs in the cartouche and the feather above it, were enameled with a blue substance let into the ivory, and the whole ground of the tablet, as well as the cartouche and part of the figures, was originally gilded, remains of the gold leaf still adhering to them. The forms and style of art have a purely Egyptian character, although there are certain peculiarities in the execution and mode of treatment that would seem to mark the work of a foreign artist. The same peculiarities, the same anomalies, characterize all the other objects discovered. Several small heads in frames, supported by pillars or pedestals most elegant in design and elaborate in execution, show not only a considerable acquaintance with art, but an intimate knowledge of the method of working in ivory. Found with them were oblong tablets upon which are sculptured with great delicacy standing figures with one hand elevated and holding in the other a stem or staff surmounted by a flower or ornament resembling the Egyptian lotus. Scattered about were fragments of winged sphinxes.

Beyond the entrance *b*, as far as chamber S, the alabaster slabs ceased altogether, and I was, for some time, at a loss to account for the manner in which the building had been continued. The pavement of baked bricks was still carried on, and it was evident that the edifice did not end here. It was some time before I discovered that I was now digging into chambers formed by walls of sun-dried bricks over which a thin coating of plaster had been laid. They had been painted with figures and ornaments, but the colors had faded so completely that scarcely any of the subjects or designs could be traced. It required the greatest care to separate the rubbish from the walls without destroying at the same time the plaster to which it adhered. I seldom succeeded in uncovering even a small portion of the paintings as the plaster fell from the walls in flakes, notwithstanding all my efforts to preserve it.

As the means at my disposal did not warrant any outlay in making more experiments without the promise of the discovery of something to carry away, I felt myself compelled, much against my inclination, to abandon the excavations in this part of the mound.

I found, by opening trenches behind chambers I and L [Northwest Palace], that similar painted rooms existed in other parts of the mound. The palace did not, therefore, only contain chambers formed by slabs of alabaster, but had other apartments.

It may be mentioned that in chamber Z I found imbedded in the wall of sun-dried bricks a small earthen bowl, or cup, of baked clay of a dark red color. This is the most ancient specimen of pottery hitherto discovered in Assyria, for from its position behind the slab it is evident that it must have been placed there at the time of the building of the edifice.

While excavating above the southern chambers of this palace [North west], I found in the rubbish several feet above the walls numerous vases of baked clay. In those that were preserved entire, human remains could be distinguished.

On the western side of the great mound, to the south of the Northwest Palace, there is a considerable elevation. To examine the place, a trench was opened. It was some time before I discovered that we were cutting into a kind of tower, or nest of upper chambers, constructed entirely of unbaked bricks, the walls being plastered and elaborately painted.

But the most important discovery connected with these upper chambers was that of the slabs forming the pavement of the two entrances. Upon them were the names and titles of five kings in genealogical succession, commencing with the father of the founder of the Northwest Palace and ending with the grandson of the builder of the center edifice [Center Palace]. By this valuable record I was able to verify the connection between the names already discovered and to add two more to the list.

In the center of the mound, on excavating to the south of the great winged bulls [Center Palace], I found a well-formed tomb built of bricks and covered with a slab of alabaster. It was about five feet in length and scarcely more than eighteen inches in breadth in the interior. On removing the lid, parts of a skeleton were exposed to view; the skull and some of the larger bones were still entire; but, on an attempt being made to move them, they crumbled into dust. With them were three earthen vessels.

Many other tombs were opened [in the center of the mound] containing vases, plates, mirrors, spoons, beads and ornaments. Some of them were built of baked bricks, carefully joined, but without mortar; others were formed by large earthen sarcophagi, covered with an entire alabaster slab.

Having carefully collected and packed the contents of the tombs, I removed them and dug deeper into the mound. I was surprised to find, about *five feet beneath them*, the remains of a building [Center Palace]. Walls of unbaked bricks could still be traced, but the slabs with which they had been cased were no longer in their places, being scattered about without order and lying mostly with their faces on the flooring of unbaked bricks. Upon them were both sculptures and inscriptions. Slab succeeded to slab, and when I had removed nearly twenty tombs and cleared away the earth from a space of about fifty feet square, the ruins which had been thus uncovered presented a very singular appearance. Above one hundred slabs were exposed to view, packed in rows, one against the other, as slabs in a stonecutter's yard or as the leaves of a gigantic book. Every slab was sculptured, and as they were placed in a regular series according to the subjects upon them, it was evident that they had been moved, in the order in which they stood, from their original positions against the walls of sun-dried brick and had been left as found preparatory to their removal elsewhere. That they were not thus arranged before being used in the building for which they had been originally sculptured was evident from the fact that the Assyrians carved their slabs after, and not before, they were placed.

These sculptures resembled, in many respects, some of the bas-reliefs found in the Southwest Palace. It appeared that the center building [Center Palace] had been destroyed to supply materials for the construction of this edifice. But here were tombs *over* the ruins. What race, then, occupied the country after the destruction of the Assyrian palaces? At what period were these tombs made? What antiquity did their presence assign to the buildings beneath them? These are questions which I am yet unable to answer and which must be left undecided until the origin and age of the contents of the tombs can be satisfactorily determined.

The bas-reliefs differed considerably from those of the Northwest Palace both in the character of the sculpture and the treatment of the subjects, in the costumes of the figures, in the caparisons of the horses and in the form of the chariots. The distinction was so great that the short period elapsing between the reigns of a father and son could scarcely have given rise, except under extraordinary circumstances, to so considerable a change in all these points. There were a few inscriptions accompanying the bas-reliefs. On the greater number of slabs, however, the space between the bas-reliefs was left without any inscription.

The subjects principally recorded by the sculptures thus found collected together, with the exception of a few gigantic figures of the king and his attendant eunuchs and of the winged priests or divinities, were battle pieces and sieges. Some cities were represented as standing on a river, in the midst of groves of date trees, others on mountains. Among the conquered people were warriors mounted on camels. It may be inferred, therefore, that a part of

British Museum

these sculptures were made to record the invasion or conquest of an Arab nation, or perhaps of a part of Babylonia, the inhabitants of the cities being assisted by auxiliaries or allies from the neighboring desert. The conquered race, as in the bas-reliefs of the Northwest Palace, were generally without helmets or armor, their hair falling loosely on their shoulders. Some, however, were provided with helmets which vary in shape from those worn by the conquerors.

A battering ram, differing in form from that seen in the earlier sculptures, was found on bas-reliefs representing sieges. They were unaccompanied by the moving tower. Some were provided with two rams, the ends of which, instead of being broad and blunt, were pointed and resembled the heads of spears.

On two slabs (occupied by one subject) were bas-reliefs of considerable interest. They represent the taking of a city, within the walls of which grew the palm and other trees. The place having been sacked, the conquerors are seen carrying away the spoil. Two eunuchs, standing near the gates, count, as they pass before them, the sheep, oxen and other cattle driven away by the warriors and write down the numbers with a pen upon rolls of paper or leather. In the lower part of the bas-relief are two carts drawn by oxen. Two women and a child are in each. The women appear to be carrying away bags containing provisions or valuable property saved during the sack. Near the gates stand two battering rams which, the city having been taken, are no longer at work. The subject is not ill arranged, and the oxen drawing the cart are well designed.

Upon other slabs were the king seated on his throne, the sun, moon and other religious symbols being placed above his head, and receiving prisoners with their arms bound behind them; eunuchs registering the heads of the enemy laid at their feet by the conquering warriors; a procession of gods borne on the shoulders of men and many other subjects.

The sides of all the slabs thus placed one against the other—the part which, in the event of their gradual covering up, would have been longest exposed—were not worn away. It was, therefore, evident that they had not been buried by the same process as the sculptures in the Northwest Palace, the walls of which could not have been long exposed. If the edifice to which they originally belonged had been suddenly covered up, it must have been subsequently excavated. The slabs were then removed from their places and arranged as they were found, preparatory to being used for other purposes, probably for the construction of the Southwest Palace. Not having been carried away, as that palace had not been finished, they were left exposed and were gradually covered by dust and rubbish. As the slabs stood on their sides and not upright, all the bas-reliefs had suffered more or less injury. Many were completely destroyed. In fact, I have little doubt from the appearance of these ruins that the building to which the sculptures originally belonged had been suddenly buried like that in the northwest corner of the mound [Northwest Palace] and that it had subsequently been uncovered, the materials being wanted for the construction, as I have conjectured, of the Southwest Palace. The slabs, not having been required, were left exposed until they were reburied by a gradual accumulation of dust and rubbish.

I have described the singular appearance presented by the ruins in the southwest corner [Southwest Palace]. Several parties of workmen were now engaged in exploring them. When all the walls still standing had been traced and trenches opened in opposite directions so that no remains of building could escape observation, I was equally at a loss to determine the position of the chambers and the extent of the edifice.

The only portion of the building sufficiently well preserved to give any idea of its original form was one large hall curiously constructed. Leading into it were two entrances formed by gigantic winged bulls and lions with human heads, and in the center was a portal formed by a second pair of bulls. At one entrance were a pair of lions with the crouching sphinxes between; at another entrance, a pair of bulls much injured, only the lower part being entire.

I will proceed to describe the walls.

All the slabs in wall *a* were unsculptured, except two. On the first were represented the walls of a castle, the king being within, seated on his throne and receiving his vizier. Around him were his attendants and above him a groom bringing corn to a horse tied to a manger. On the other slab was a horseman wearing a helmet with a curved crest. He appears to be raising his hand in the act of asking for quarter, while his horse, pierced by the spears of two pursuing warriors, is rearing and plunging. Both the slabs had been greatly injured.

On the slab adjoining entrance *e* were two bas-reliefs, the upper (partly destroyed) representing warriors hewing down trees, the lower, a warrior on horseback hunting the wild bull. Both were too much injured to bear removal.

On a prostrate slab were two bas-reliefs. The upper was so much injured

British Museum

that the outlines of a chariot and warriors on foot could with difficulty be traced. The lower was the siege of a castle; a eunuch was represented discharging his arrows against warriors without helmets who manned the towers and walls. The besiegers were leading away prisoners and carrying off the spoil. One high-capped warrior was seen cutting a bucket from a rope passed through a pulley, and probably used by the besieged to supply themselves with water from a well without the castle walls. The pulley resembled those now in common use for raising and lowering buckets into wells. This bas-relief had been brought from the Northwest Palace.

Upon the three slabs forming the wall *r* were bas-reliefs of considerable interest. They once formed part of another building but do not belong to either the Northwest or the Center Palace. In the lower compartment of the first one was a charioteer in a highly ornamented chariot, the horse being held by a groom on foot preceded by a eunuch.

On the lower part of the second was the king placing his foot on the neck of a prostrate prisoner and raising his spear over him. Following the king is a eunuch carrying a fan, and standing before him his vizier, also attended by a eunuch.

On the lower part of the third slab was a bas-relief of considerable interest. It represented either a procession of gods borne on the shoulders of warriors or warriors returning from the sack of a city, carrying away the idols of the conquered people. Each figure was raised by four men; the first was that of a female seated on a high-backed armchair looking toward the spectator; the face is, consequently, sculptured in full—a rare occurrence in Assyrian sculpture.

As the *bottom* of the slabs forming this edifice was even above the level of the *top* of those in the Northwest Palace, and as no building had yet been found from which many of the sculptures could have been taken, it appeared to me possible that the Southwest Palace stood above other ruins. By way of experiment, I directed long and very deep trenches to be opened in three different directions; nothing, however, was discovered but a box or square hole, formed by bricks carefully fitted together, containing several small heads in unbaked clay of a dark brown color. These heads are furnished with beards and have very high pointed caps (not helmets) or miters. They were found about twenty feet beneath the surface and were probably idols placed for some religious purpose under the foundations of buildings. Objects somewhat similar, in unbaked clay, were discovered at Khorsabad.

The southeast corner of the mound, which was considerably higher than any other part, appears to have been the principal burying place of those who occupied the country after the destruction of the oldest of the Assyrian palaces. The sarcophagi were mostly of the shape of a dish cover, but there were other tombs constructed of bricks, well fitted together and covered by a slab, similar to those above the ruins of the edifice in the center of the mound [Center Palace]. Removing these tombs I discovered beneath them the remains of a building and explored seven chambers. No sculptured slabs or inscriptions were found in them.

Several trenches were opened in other parts of the mound. Everywhere I found traces of buildings and generally reached a pavement of baked bricks between ten and fifteen feet beneath the surface.

The ruins were, of course, very inadequately explored, but with the small sum at my disposal I was unable to pursue my researches to the extent that I could have wished. If, after carrying a trench to a reasonable depth and distance, no remains of sculpture or inscription appeared, I abandoned it and renewed the experiment elsewhere. By this mode of proceeding I could ascertain, at least, that in no part of the mound was there any very extensive edifice still standing. I have left a great part of the mound of Nimroud to be explored by those who may hereafter succeed me in the examination of the ruins of Assyria.

Chapter 7

■ Assyria proper, like Babylonia, owed its ancient fertility as much to the system of artificial irrigation, so extensively and successfully adopted by the inhabitants of the country, as to the rains which fell during the winter and early spring. The Tigris and Euphrates, unlike the Nile, did not overflow their banks and deposit a rich manure on the face of the land. They rose sufficiently at the time of the melting of the snows in the Armenian hills to fill the numerous canals led from them into the adjacent country, but their beds were generally so deep, or their banks so high, that when the stream returned to its usual level, water could only be raised by artificial means.

Rain, amply sufficient to insure the most plentiful crops, generally falls during the winter. Indeed, such is the richness of the soil of Assyria that even a few heavy showers in the course of the year, at the time of sowing the seed and when the corn is about a foot above the ground, are sufficient to insure a good harvest. It frequently, however, happens that the season passes without rain. Such was the case this year [1847]. During the winter and spring, no water fell. The inhabitants of the villages had put their whole stock of wheat and barley into the ground. They now looked in despair upon the cloudless sky. I watched the young grass as it struggled to break through the parched earth, but it was burned up almost at its birth. Sometimes a distant cloud hanging over the solitary hill of Arbela or rising from the desert in the far west led to hopes, and a few drops of rain gave rise to general rejoicings. The Arabs would then form a dance and raise songs and shouts, the women joining. But disappointment always ensued. The clouds passed over, and the same pure blue sky was above us. The total absence of verdure in spring was painful. For months my eye had not rested upon a green thing, and that unchanging yellow barren waste has a depressing effect upon the spirits. I had been looking forward to the return of the grass to encamp outside Mosul and had meditated many excursions to ancient ruins in the desert and the mountains, but I was doomed to disappointment.

The Pasha issued orders that Christians as well as Moslems should join in a general fast and in prayers. Supplications were offered up in the churches and the mosques. If a cloud were seen on the horizon the inhabitants of the village would immediately walk into the open country to chant prayers and verses from the Koran. Crazy ascetics who wandered over the country, either half clothed in the skins of lions or gazelles or stark naked, burned

The Metropolitan Museum of Art,
Gift of John D. Rockefeller, Jr., 1931 (detail)

themselves with hot irons and ran shouting about the streets of Mosul.

There was not even the prospect of a shower. A famine appeared to be inevitable. It was known, however, that there were abundant supplies of corn in the granaries of the principal families of Mosul, and the fact having been brought to the notice of the Pasha, he at once ordered the stores to be opened and their contents to be offered for sale in the market at moderate prices. As usual, the orders were given to the very persons who were speculating upon the miseries of the poor and needy—to the cadi, the mufti and the head people of the town. They proceeded to obey, with great zeal and punctuality, the orders of his Excellency, but somehow or another overlooked their own stores and those of their friends and ransacked the houses of the rest of the inhabitants. In a few days, consequently, those who had saved up a little grain for their own immediate wants were added to the number of the starving, and the necessities and misery of the town were increased.

The Bedouins, who are dependent upon the villages for supplies, now also began to feel the effects of the failure of the crops. As is generally the case in such times, they were preparing to make up for their sufferings by plundering the caravans of merchants and the peaceable inhabitants of the districts within reach of the desert. Casual plundering parties made their appearance among the villages, and it was predicted that as soon as their tents were pitched nearer the town the country would be not only very unsafe but almost uninhabitable.

These circumstances induced me to undertake the removal of the larger sculptures from Nimroud as early as possible. The dry season had enabled me to carry on the excavations without interruption. As no fain fell to loosen the earth above the ruins there was no occasion to prop up the sides of the trenches or to cover the sculptures; considerable expense was thus saved. Had there been the usual violent storms, not only would the soil have continually fallen in and reburied the building, but the bas-reliefs would have been exposed to injury. A marsh would also have been formed round the base of the mound, completely cutting me off from the river and impassable to any cart carrying the larger sculptures.

This year, however, there was no marsh round the ruins. I determined, therefore, to send the sculptures to Busrah in the month of March or April, foreseeing that as soon as the Bedouins had moved northward from Babylonia and had commenced their plundering expeditions in the vicinity of Mosul, I should be compelled to leave Nimroud.

The Trustees of the British Museum had not contemplated the removal of either a winged bull or lion, and I had at first believed that with the means at my disposal it would have been useless to attempt it. They wisely determined that these sculptures should not be sawn into pieces to be put together again in Europe, as the pair of bulls from Khorsabad. They were to remain where discovered until some favorable opportunity of moving them entire might occur, and I was directed to heap earth over them after the excavations had been brought to an end.

Being loath, however, to leave all these fine specimens of Assyrian sculpture behind me, I resolved upon attempting the removal and embarkation of two of the smallest and best preserved. Those fixed upon were a lion from hall Y in the Northwest Palace and a bull from the same hall.

Thirteen pairs of these gigantic sculptures and several fragments of others had been discovered, but many of them were too much injured to be worth moving. I had wished to secure the pair of lions forming the great entrance (*a*) into the principal chamber (B) of the Northwest Palace, the finest specimen of Assyrian sculpture discovered in the ruins. But after some deliberation I determined to leave them for the present, as, from their size, the expense attending their conveyance to the river would have been very considerable.

I formed various plans for lowering the smaller lion and bull, for dragging them to the river and for placing them upon rafts. I resolved upon constructing a cart sufficiently strong to bear any of the masses to be moved. To enable me to move the bull from the ruins and to place it on the cart in the plain below, it was necessary to cut a trench nearly two hundred feet long, about fifteen feet wide and, in some places, twenty feet deep. A road was thus constructed from the entrance in which stood the bull to the edge of the mound. There being no means at my disposal to raise the sculpture out of the trenches, like the smaller bas-reliefs, this road was necessary. It was a tedious undertaking, as a very large accumulation of earth had to be removed. About fifty Arabs and Nestorians were employed in the work.

On opening the trench it was found that a chamber had once existed to the west of hall Y [Northwest Palace]. The sculptured slabs forming its sides had been destroyed or carried away. Part of the walls of unbaked bricks, however, could still be traced. The only bas-relief discovered was lying flat on the pavements where it had evidently been left when the adjoining slabs were removed. It represents a lion hunt. Only one lion, wounded

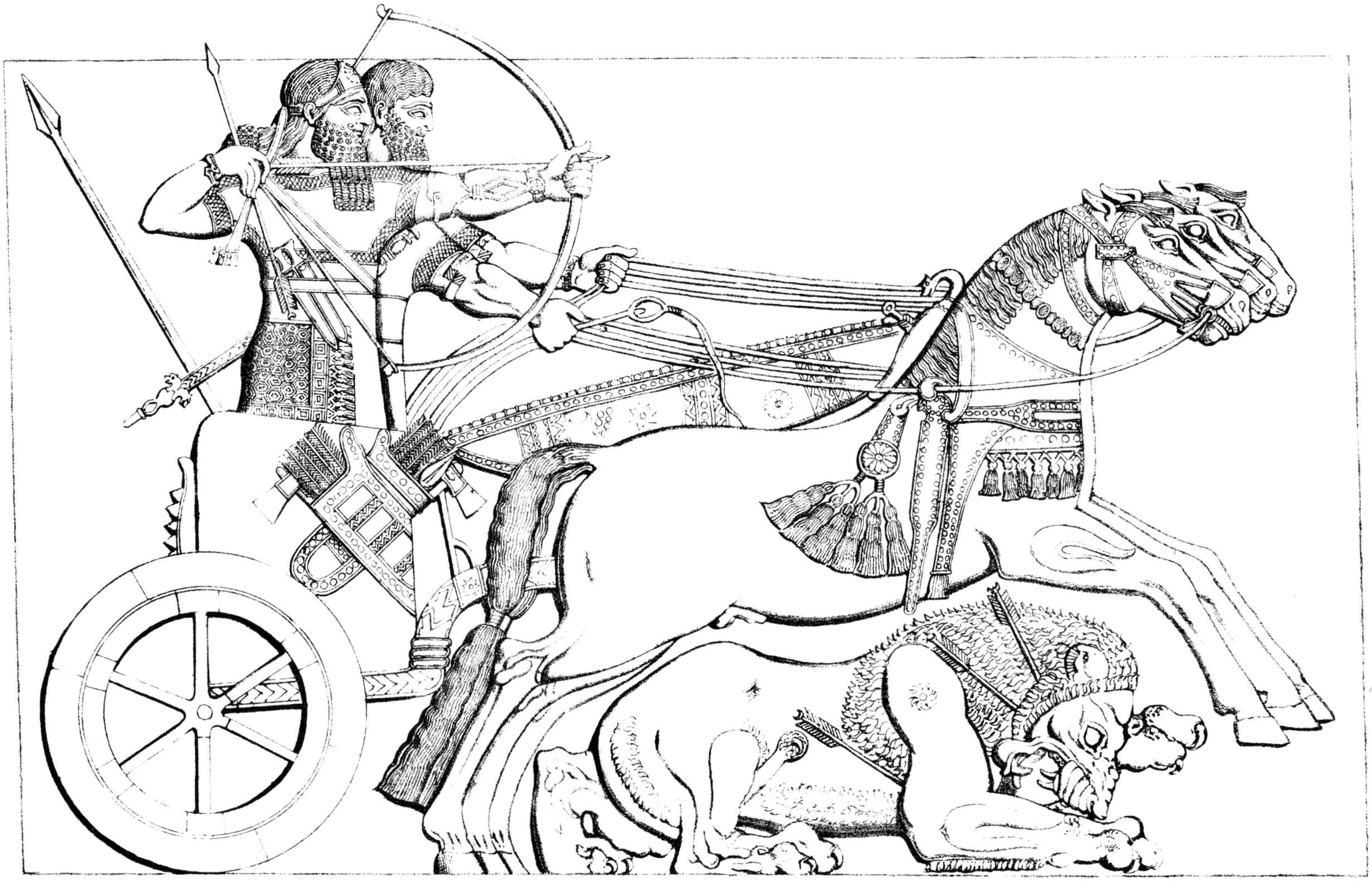

From Layard,
Nineveh and Its Remains

and under the horse's feet, is visible. A warrior in a chariot is discharging his arrows at some object before him. It is evident that the subject must have been continued on an adjoining slab on which was probably represented the king joining in the chase. This small bas-relief is remarkable for its finish, the elegance of the ornaments and the great spirit of the design. In these respects it resembles the battle scene in the Southwest Palace, and I am inclined to believe that they both belonged to this ruined chamber in which, perhaps, the sculptures were more elaborate and more highly finished than in any other part of the building.

The work of different artists may be plainly traced in the Assyrian edifices. Frequently where the outline is spirited and correct and the ornaments designed with considerable taste, the execution is defective or coarse, evidently showing that while the subject was drawn by a master the carving of the stone had been entrusted to an inferior workman. In many sculptures some parts are more highly finished than others, as if they had been retouched by an experienced sculptor. The figures of the enemy are generally rudely drawn and left unfinished, to show, probably, that being those of the conquered or captive race they were unworthy the care of the artist. It is rare to find an entire bas-relief equally well executed in all its parts.

As the bull was to be lowered on its back, the unsculptured side of the slab having to be placed on rollers, I removed the walls behind it. An open space was thus formed, large enough to admit of the sculpture when prostrate and leaving room for the workmen to pass on all sides of it. The principal difficulty was of course to lower the mass. When once on the ground or on rollers it could be dragged forward by the united force of a number of men, but during its descent it could only be sustained by ropes. If, not strong enough to bear the weight, they chanced to break, the sculpture would be precipitated to the ground and would probably be broken in the fall. The few ropes I possessed had been expressly sent to me across the desert from Aleppo, but they were small. From Baghdad I had obtained a thick hawser made of the fibers of the palm. In addition I had been furnished with two pairs of blocks and a pair of jackscrews belonging to the steamers of the Euphrates expedition. These were all the means at my command for moving the bull and lion. The sculptures were wrapped in mats and felts to preserve them as far as possible from injury in case of a fall and to prevent the ropes chipping or rubbing the alabaster.

The bull was ready to be moved by the eighteenth of March. The earth had been taken from under it, and it was now only supported by beams resting against the opposite wall. Several thick rollers were placed upon sleepers, or half beams, formed out of the trunks of poplar trees, well greased and laid on the ground parallel to the sculpture. The bull was to be lowered upon these rollers. A deep trench had been cut behind the second bull completely across the wall and, consequently, extending from chamber to chamber. A bundle of ropes coiled round this isolated mass of earth served to hold two blocks, two others being attached to ropes wound round the bull to be moved. The ropes by which the sculpture was to be lowered were passed through these blocks, the ends of the tackle, as they are technically called, being led from the blocks above the second bull and held by the Arabs. The cable having been first passed through the trench and then round the sculpture, the ends were given to two bodies of men. Several of the strongest

Chaldeans placed thick beams against the back of the bull and were directed to withdraw them gradually, supporting the weight of the slab and checking it in its descent in case the ropes should give away.

The inhabitants of Naifa and Nimroud, having volunteered to assist on the occasion, were distributed among my Arabs. The workmen, except the Chaldeans who supported the beams, were divided into four parties, two of which were stationed in front of the bull and held the ropes passed through the blocks. The rest clung to the ends of the cable and were directed to slack off gradually as the sculpture descended.

The men being ready and all my preparations complete, I stationed myself on the top of the high bank of earth over the second bull and ordered the wedges to be struck out from under the sculpture to be moved. Still, however, it remained firmly in its place. A rope having been passed round it, six or seven men easily tilted it over. The thick, ill-made cable stretched with the

Lowering the great winged bull. From Layard, Nineveh and Its Remains

strain and almost buried itself in the earth round which it was coiled. The ropes held well. The mass descended gradually, the Chaldeans propping it up firmly with the beams. It was a moment of great anxiety. The drums and shrill pipes of the Kurdish musicians increased the din and confusion caused by the war cry of the Arabs, who were half frantic with excitement. They had thrown off nearly all their garments, their long hair floated in the wind and they indulged in the wildest postures and gesticulations as they clung to the ropes. The women had congregated on the sides of the trenches and by their incessant screams added to the enthusiasm of the men. The bull once in motion, it was no longer possible to be heard. The loudest cries I could produce were buried in the heap of discordant sounds. Neither the hippopotamus-hide whips of the cawasses nor the bricks and clods of earth with which I endeavored to draw attention from some of the most noisy of the group were of any avail. Away went the bull, steady enough as long as supported by the props behind, but as it came nearer to the rollers the beams could no longer be used. The cable and ropes stretched more and more. Dry from the climate, as they felt the strain they creaked and threw out dust. Water was thrown over them, but in vain, for they all broke together when the sculpture was within four or five feet of the rollers. The bull was precipitated to the ground. Those who held the ropes, thus suddenly released, followed its example and were rolling one over the other in the dust. A sudden silence succeeded to the clamor. I rushed into the trenches, prepared to find the bull in many pieces. It would be difficult to describe my satisfaction when I saw it lying precisely where I had wished to place it, and uninjured!

The Arabs no sooner got on their legs again than seeing the result of the accident, they darted out of the trenches and, seizing by the hands the women who were looking on, formed a large circle and, yelling their war cry with redoubled energy, commenced a most mad dance. The musicians exerted themselves to the utmost, but their music was drowned by the cries of the dancers. It would have been useless to endeavor to put any check upon these proceedings. I preferred allowing the men to wear themselves out—a result which, considering the amount of exertion and energy displayed both by limbs and throat, was not long in taking place.

I now prepared with the aid of the Bairakdar and the Tiyari [Nestorian Christian Chaldeans] to move the bull into the long trench which led to the edge of the mound. As soon as the excitement of the Arabs had sufficiently abated to enable them to resume work, the sculpture was dragged out of its place by ropes.

The sun was going down; I deferred any further labor to the morrow. The Arabs dressed themselves and, placing the musicians at their head, marched toward the village singing their war songs and occasionally raising a wild yell, throwing their lances into the air and flourishing their swords and shields over their heads.

Everything had been prepared on the previous day for moving the bull, and the men had now only to haul on the ropes. There was little difficulty in dragging it down the precipitous side of the mound. The bull itself was then lowered onto a cart by still further digging away the soil. It was soon ready to be dragged to the river. The procession was formed. I rode first, with the Bairakdar, to point out the road. Then came the musicians with their drums and fifes, drumming and fifing with might and main. The cart followed, dragged by about three hundred men, all screeching at the top of their voices and urged on by the cawasses and superintendents. The procession was closed by the women, who kept up the enthusiasm of the Arabs by their shrill cries.

We advanced well enough, although the ground was very heavy, until we reached the ruins of the former village of Nimroud. It is the custom in this part of Turkey for the villagers to dig deep pits to store their corn, barley and straw for the autumn and winter. These pits generally surround the villages. Although I had carefully examined the ground before starting, one of these holes had escaped my notice, and into it two wheels of the cart completely sank. The Arabs pulled and yelled in vain. The ropes broke, but the wheels refused to move. We tried every means to release them, but unsuccessfully. After working until dusk, we were obliged to give up the attempt.

Next morning we succeeded in clearing away the earth and in placing thick planks beneath the buried wheels. After a few efforts the cart moved forward amid the shouts of the Arabs who, as was invariably their custom on such occasions, indulged, while pulling at the ropes, in the most outrageous antics. The procession was formed as on the previous day, and we dragged the bull triumphantly down to within a few hundred yards of the river. Here the wheels buried themselves in the sand, and it was night before we contrived, with the aid of planks and by increased exertions, to place the sculpture on the platform prepared to receive it and from which it was to slide down on the raft. The tents of the Arabs, who encamped near the river, were pitched round the bull until its companion, the lion, should be brought down, and the two embarked together for Baghdad. The night was passed in renewed

Procession of the bull beneath the mound of Nimrud. From Layard, A Popular Account of Discoveries at Nineveh. *Photograph by Alfred Tamarin*

rejoicings to celebrate the successful termination of our labors. On the following morning I rode to Mosul to enjoy a few days' rest after my exertions.

In the middle of April I determined to lower the lion at once on the cart and not to drag it out of the mound over the rollers. This sculpture, during its descent, was supported in the same manner as the bull had been; but to avoid a second accident I doubled the number of ropes and coils of the cable. Enough earth was removed to bring the top of the cart to a level with the bottom of the lion.

As the lion was cracked in more than one place, considerable care was required in lowering and moving it. Both, however, were effected without accident. The Arabs assembled as they had done at the removal of the bull. We had the same shouting and the same festivities. The lion descended into the place I had prepared for it on the cart and was easily dragged out of the ruins. It was two days in reaching the river, as the wheels of the cart sank more than once into the loose soil and were with difficulty extricated.

The lion and bull were at length placed side by side on the banks of the Tigris, ready to proceed to Busrah as soon as I could make the necessary arrangements for embarking them on rafts.

The sculptures, which I had hitherto sent to Busrah, had been floated down the river on rafts as far only as Baghdad. There they had been placed in boats built by the natives for the navigation of the lower part of the Tigris and Euphrates. These vessels, principally constructed of thin poplar planks, reeds and bitumen, were much too small and weak to carry either the lion or the bull and indeed, had they been large enough, it would have been difficult if not impossible in the absence of proper machinery to move such heavy masses into them. I resolved, therefore, to attempt the navigation of the lower, as well as of the upper, part of the river with rafts, and to embark the lion and bull at once for Busrah. The raftmen of Mosul, who are accustomed to navigate the Tigris to Baghdad but never venture farther, pronounced the scheme to be impracticable and refused to attempt it.

It was impossible by the most convincing arguments, even though supported by the exhibition of a heap of coins, to prevail upon the raftmen of Mosul to construct such rafts as I required or to undertake the voyage. I therefore found a man at Baghdad who declared himself willing to make the attempt.

The beams of poplar wood, forming an inclined plane from beneath the sculptures to the rafts, were first well greased. A raft supported by six hundred skins having been brought to the river bank opposite the bull, the wedges were removed from under the sculpture which immediately slid down into its place. The only difficulty was to prevent its descending too rapidly and bursting the skins by the sudden pressure. The Arabs checked it by ropes, and it was placed without any accident. The lion was then embarked in the same way upon a second raft of the same size.

Many slabs, including the large bas-reliefs of the king on his throne between the eunuchs and winged figures and more than thirty cases containing small objects discovered in the ruins, were placed on the rafts with the lion and bull.

On the morning of the twenty-second, all the sculptures having been embarked, I gave two sheep to the raftmen to be slain on the bank of the river as a sacrifice to insure the success of the undertaking. The carcasses were distributed, as is proper on such occasions, among the poor. A third sheep was reserved for a propitiatory offering. All ceremonies having been duly performed, the rafts slowly floated, with the cargo charge, down the stream. I watched the rafts until they disappeared behind a projecting bank forming a distant reach of the river. I could not forbear musing upon the strange destiny of their burdens which, after adorning the palaces of the Assyrian kings, the objects of the wonder and maybe the worship of thousands, had been buried unknown for centuries beneath a soil trodden by Persians under Cyrus, by Greeks under Alexander and by Arabs under the first descendants of their prophet. They were now to visit India, to cross the most distant seas of the southern hemisphere and to be finally placed in a British museum.

After the departure of the lion and bull, I now commenced burying in those parts of the ruins which still remained exposed, according to the instructions I had received from the Trustees of the British Museum. Had the numerous sculptures been left, without any precaution being taken to preserve them, they would have suffered not only from the effects of the atmosphere but from the spears and clubs of the Arabs, who are always ready to knock out the eyes and to otherwise disfigure the idols of the unbelievers. The

Moving the winged lion in the British Museum. From Joseph Bonomi, Nineveh and Its Palaces. *Photograph by Alfred Tamarin*

rubbish and earth removed on opening the building was accordingly brought back in baskets, thrown into the chambers and heaped over the slabs until the whole was again covered over.

But before leaving Nimroud and reburying its palaces, I would wish to lead the reader once more through the ruins of the principal edifice [Northwest Palace] and to convey as distinct an idea as I am able of the excavated halls and chambers as they appeared when fully explored. Let us imagine ourselves issuing from my tent near the village in the plain. On approaching the mound, not a trace of building can be perceived except a small mud hut covered with reeds, erected for the accommodation of my Chaldean workmen. We ascend this artificial hill, but still see no ruins, not a stone protruding from the soil. There is only a broad level platform before us, perhaps covered with a luxuriant crop of barley, or maybe yellow and parched, without a blade of vegetation except here and there a scanty tuft of camel thorn. Low black heaps, surrounded by brushwood and dried grass, a thin column of smoke issuing from the midst of them, may be seen here and there. These are the tents of the Arabs, and a few miserable old women are groping about them, picking up camel's dung or dry twigs. One or two girls, with firm step and erect carriage, are perceived just reaching the top of the mound with the water jar on their shoulders or a bundle of brushwood on their heads. On all sides of us, apparently issuing from underground, are long lines of wild-looking beings with disheveled hair, their limbs only half concealed by a short loose shirt, some jumping and capering and all hurrying to and fro, shouting like madmen. Each one carries a basket and as he reaches the edge of the mound, or some convenient spot near, empties its contents, raising at the same time a cloud of dust. He then returns at the top of his speed, dancing and yelling as before and flourishing his basket over his head; again he suddenly disappears in the bowels of the earth from whence he emerged. These are the workmen employed in removing the rubbish from the ruins.

We will descend into the principal trench by a flight of steps rudely cut into the earth near the western face of the mound. As we approach it we find a party of Arabs bending on their knees and intently gazing at something beneath them. Each holds his long spear tufted with ostrich feathers in one hand, and in the other the halter of his mare, which stands patiently behind him. The party consists of a Bedouin Sheikh from the desert and his followers who, having heard strange reports of the wonders of Nimroud, have made several days' journey to remove their doubts and satisfy their curiosity.

We descend about twenty feet into the trenches and suddenly find ourselves between a pair of colossal lions, winged and human-headed, forming a portal.

In the subterranean labyrinth which we have reached, all is bustle and confusion. Arabs are running about in different directions, some bearing baskets filled with earth, others carrying the water jars to their companions. The Chaldeans or Tiyari, in their striped dresses and curious conical caps, are digging with picks into the tenacious earth, raising a dense cloud of fine dust at every stroke. The wild strains of Kurdish music may be heard occasionally issuing from some distant part of the ruins and, if they are caught by the parties at work, the Arabs join their voices in chorus, raise the war cry and labor with renewed energy. Leaving behind us a small chamber in which the sculptures are distinguished by a want of finish in the execution and considerable rudeness in the design of the ornaments, we issue from between the winged lions and enter the remains of the principal hall. On both sides of us are sculptured gigantic, winged figures, some with the heads of eagles, others entirely human and carrying mysterious symbols in their hands. To the left is another portal, also formed by winged lions. One of them has, however, fallen across the entrance, and there is just room to creep beneath it. Beyond this portal is a winged figure and two slabs with bas-reliefs, but they have been so much injured that we can scarcely trace the subject upon them. Farther on there are no traces of wall, although a deep trench has been opened. The opposite side of the hall has also disappeared, and we only see a high wall of earth. On examining it attentively, we can detect the marks of masonry, and we soon find that it is a solid structure built of bricks of unbaked clay, now of the same color as the surrounding soil and scarcely to be distinguished from it.

The slabs of alabaster, fallen from their original position, have, however, been raised, and we tread in the midst of a maze of small bas-reliefs representing chariots, horsemen, battles and sieges.

Having walked about one hundred feet among these scattered monuments of ancient history and art, we reach another doorway formed by gigantic winged bulls in yellow limestone. One is still entire, but its companion has fallen and is broken into several pieces—the great human head is at our feet.

We pass on without turning into the part of the building to which this portal leads. Beyond it we see another winged figure holding a graceful flower in its hand and apparently presenting it as an offering to the winged

bull. Adjoining this sculpture we find eight fine bas-reliefs. There is the king, hunting and triumphing over the lion and wild bull, and the siege of the castle with the battering ram. We have now reached the end of the hall and find before us an elaborate and beautiful sculpture representing two kings standing beneath the emblem of the supreme deity and attended by winged figures. Between them is the sacred tree. In front of this bas-relief is the great stone platform upon which, in days of old, may have been placed the throne of the Assyrian monarch when he received his captive enemies or his courtiers.

To the left of us is a fourth outlet from the hall, formed by another pair of lions. We issue from between them and find ourselves on the edge of a deep ravine, to the north of which rises, high above us, the lofty ziggurat. Figures of captives bearing objects of tribute—earrings, bracelets and monkeys—may be seen on walls near this ravine, and two enormous bulls and two winged figures above fourteen feet high are lying on its very edge.

As the ravine bounds the ruins on this side, we must return to the yellow bulls. Passing through the entrance formed by them, we enter a large chamber surrounded by eagle-headed figures: at one end of it is a doorway guarded by two priests or divinities and in the center another portal with winged bulls. Whichever way we turn we find ourselves in the midst of a nest of rooms and without an acquaintance with the intricacies of the place we should soon lose ourselves in this labyrinth. The accumulated rubbish being generally left in the center of the chambers, the whole excavation consists of a number of narrow passages, paneled on one side with slabs of alabaster and shut in on the other by a high wall of earth, half buried, in which may here and there be seen a broken vase or a brick painted with brilliant colors. We may wander through these galleries for an hour or two, examining the marvelous sculptures or the numerous inscriptions that surround us. Here we meet long rows of kings attended by their eunuchs and priests, there lines of winged figures carrying fir cones and religious emblems and seemingly in adoration before the mystic tree. Other entrances, formed by winged lions and bulls, lead us into new chambers. In every one of them are fresh objects of curiosity and surprise. At length wearied, we issue from the buried edifice by a trench on the opposite side to that by which we entered and find ourselves again upon the naked platform. We look around in vain for any traces of the wonderful remains we have just seen and are half inclined to believe that we have dreamed a dream or have been listening to some tale of Eastern romance.

Some, who may hereafter tread on the spot when the grass again grows over the ruins of the Assyrian palaces, may indeed suspect that I have been relating a vision.

Winged bull in the Metropolitan Museum of Art. Photograph by Alfred Tamarin

Chapter 8

■ The chambers at Nimroud had been filled up with earth, and the sculptures thus preserved from injury. The surrounding country became daily more dangerous from the incursions of the Arabs of the desert, who now began to encamp even on the west bank of the Tigris. It was time, therefore, to leave the village.

As a small sum of money still remained at my disposal, I proposed to devote it to an examination of the ruins opposite Mosul, particularly of the great mound of Kouyunjik. Although excavations on a small scale had already been made there, I had not hitherto had time to superintend them myself, and in such researches, the natives of the country cannot be trusted.

It is well known that almost since the fall of the Assyrian Empire, a city of some extent, representing the ancient Nineveh, has stood on the banks of the Tigris in this part of its course. The modern city may not have been built above the ruins of the ancient, but it certainly arose in their immediate vicinity. The alabaster slabs which had once lined the walls of the old palaces and still remained concealed by the earth forming the mound had been frequently exposed by accident or by design. Those who were settling in the neighborhood soon found that the ruins offered an inexhaustible mine of building materials. The alabaster was dug out to be either used entire in the construction of houses or to be burned for lime.

A few years ago a bas-relief had been discovered in one part of the ruins during a search after stones for the erection of a bridge across the Tigris. The removal of slabs and the destruction of sculptures for similar purposes may have been going on for centuries. There was, therefore, good reason to doubt whether any edifice, even in an imperfect state, still existed in Kouyunjik. I knew that under the village containing the tomb of the prophet Jonah there were remains of considerable importance, probably as entire as those discovered at Nimroud. They owe their preservation to the existence, from a very remote period, of the tomb and village above them. Portions of sculpture and inscriptions had frequently been found when the inhabitants of the place had dug the foundations of their dwellings. But the prejudices of the people of Mosul forbade any attempt to explore a spot so venerated for its sanctity.

In excavating at Kouyunjik, I pursued the plan I had adopted at Nimroud. I resided in the town. The Arabs pitched their tents on the summit of the mound at the entrances to the trenches. The Tiyari encamped at its foot,

The Metropolitan Museum of Art, Gift of John D. Rockefeller, Jr., 1931 (detail)

on the banks of the Khausser, the small stream which flows through the ruins. Here the men and women found a convenient place for their constant ablutions. The nearness of the ruins to Mosul enabled the inhabitants of the town to gratify their curiosity by a constant inspection of my proceedings, and a great crowd of gaping Moslems and Christians was continually gathered round the trenches. I rode to the mound early every morning and remained there during the day.

An acquaintance with the nature and position of the ancient edifices of Assyria will at once suggest the proper method of examining the mounds which enclose them. The Assyrians, when about to build a palace or public edifice, appear to have first constructed a platform, or solid compact mass of sun-dried bricks, about thirty or forty feet above the level of the plain. Upon it they raised the monument. When the building was destroyed, its ruins, already half buried by the falling in of the upper walls and roof, remained of course on the platform and were in time completely covered up by the dust and sand carried about by the hot winds of summer. Consequently, in digging for remains, the first step is to reach the platform of sun-dried bricks. When this is discovered, the trenches must be opened to the level of it and not deeper; they should then be continued in opposite directions, care being always taken to keep along the platform. By this means, if there be any ruins, they must necessarily be discovered, supposing the trenches to be long enough, for the chambers of the Assyrian edifices are generally narrow, and their walls, or the slabs which cased them if fallen, must sooner or later be reached.

At Kouyunjik, the accumulation of rubbish and earth was considerable, and trenches were dug to the depth of twenty feet before the platform of unbaked bricks was discovered. Before beginning the excavations, I carefully examined all parts of the mound to ascertain where remains of buildings might most probably exist and at length decided upon continuing my researches where I had commenced them last summer, near the southwest corner.

The workmen had been digging for several days without finding any other remains than fragments of calcined alabaster, sufficient, however, to encourage me to persevere in the examination of this part of the ruins. One morning as I was in Mosul, two Arab women came to me and announced that sculptures had been discovered. They had hurried from the mounds as soon as the first slab had been exposed to view and blowing up the skins which they always carry about with them had swum upon them across the river. They had scarcely received the present claimed in the East by the bearers of good tidings, than one of my overseers, who was generally known from his corpulence as Toma Shishman, or fat Toma, made his appearance, breathless from his exertions. He had hurried as fast as his legs could carry him over the bridge to obtain the reward carried off, in this instance, by the women.

I rode immediately to the ruins and, on entering the trenches, found that the workmen had reached a wall and the remains of an entrance. The only slab as yet uncovered had been almost completely destroyed by fire. It stood on the edge of a deep ravine which ran far into the southern side of the mound.

As the excavations at Kouyunjik were carried on in precisely the same manner as those at Nimroud, I need not trouble the reader with any detailed account of my proceedings. The wall first discovered proved to be the side of a chamber. By following it we reached an entrance formed by winged bulls, leading into a second hall. In a month nine chambers had been explored.

The palace had been destroyed by fire. The alabaster slabs were almost reduced to lime and many of them fell to pieces as soon as uncovered. The places, which others had occupied, could only be traced by a thin white deposit left by the burnt alabaster upon the wall of sun-dried bricks and having the appearance of a coating of plaster.

In its architecture the newly discovered edifice resembled the palaces of Nimroud and Khorsabad. The chambers were long and narrow. The walls were of unbaked bricks with a paneling of sculptured slabs. The bas-reliefs were, however, much larger in their dimensions than those generally found at Nimroud, being about ten feet high and from eight to nine feet wide. The winged human-headed bulls forming the entrances were from fourteen to sixteen feet square. The slabs, unlike those I had hitherto discovered, were not divided in the center by a band of inscription, but were completely covered with figures. In the costumes of the warriors and in the trappings and caparisons of the horses, the sculptures were similar to those of Khorsabad.

The name of the king, occurring both on the backs of the slabs and on the bricks, resembles that occupying the second place in the genealogical list in the short inscriptions on the bulls and lions of the southern, or most recent, palace of Nimroud. He was the son of the builder of Khorsabad. The comparative epoch of the foundation of this edifice can thus be ascertained.

In the earth above the edifice at Kouyunjik a few earthen vases and fragments of pottery were discovered, but no sarcophagi or tombs with human remains like those of Nimroud and Kalah Sherghat. The foundations of buildings of roughly hewn stone were also found above the Assyrian ruins.

I will describe the subjects of some of the sculptures that could still be traced on the walls of the chambers.

A spacious entrance at the upper end of the great hall [B] opened into a small chamber.

Upon the two slabs beyond this entrance was a subject of considerable interest. A man stood at the castle gate which opened immediately upon the water. A woman, who had already embarked in one of the ships, was seen stretching out her arms to receive a child which the man was giving to her. The sea was indicated by wavy lines carried across the slab from top to bottom and by fish, crabs and turtles. The vessels were of two kinds: some had masts and sails as well as oars, others were impelled by rowers alone. They were furnished with two decks. On the upper stood warriors armed with spears and women wearing high turbans or miters. On the lower were double sets of rowers—eight and sometimes ten men sitting on a side, making sixteen or twenty in all. The sides of the upper deck as well as the battlements of the castle on the seashore were hung with shields.

The entrance *b* of chamber C, formed by two winged bulls, led into a further chamber, a part only of which I was able to explore. The alabaster slabs had in many places completely disappeared, and I could not even trace the walls and form of the apartment. On the first and second slabs was portrayed a mountainous country; a river ran through the midst of it.

The remaining slabs were covered from top to bottom with rows of warriors, spearmen and archers in their respective costumes and in military array. Each slab must have contained several hundred minute figures sculptured with great care and delicacy.

On the bottom of one slab was a fisherman fishing with a hook and line in a pond. Upon his back was a wicker basket containing the fish he had caught. This was almost the only fragment of sculpture that I was able to move and send to England as a specimen of the bas-reliefs of Kouyunjik.

The bas-reliefs on the walls of chamber F recorded the victories and triumphs of the king in a mountainous country and the siege of a city standing on the banks of a river. The king stood in his chariot, and around him were warriors leading away horses and captives, bringing heads to the scribes and contending in battle with the enemy. All the slabs, however, were equally

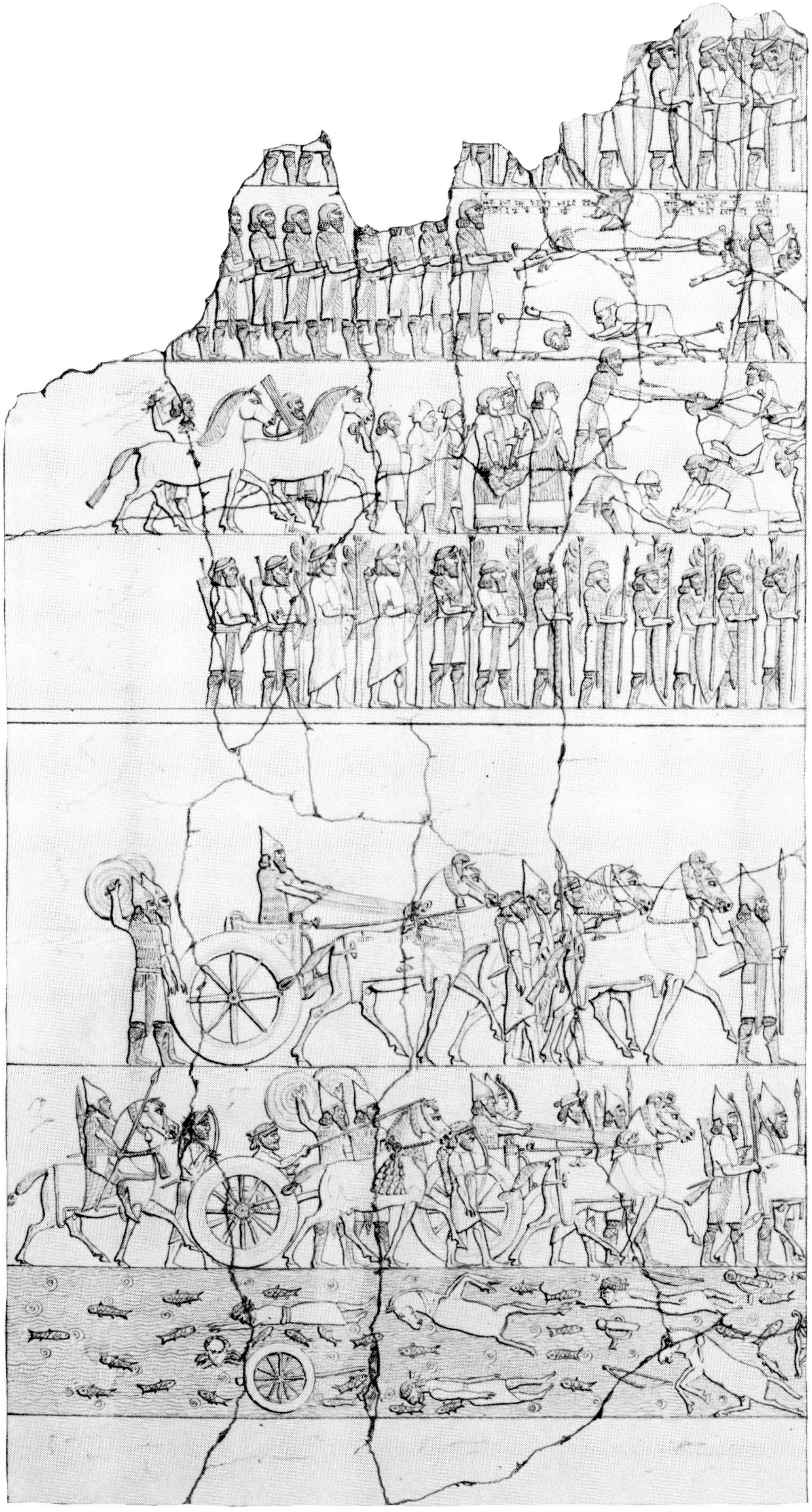

British Museum

injured. The figures upon them could scarcely be distinguished.

The wide portal formed by the winged bulls at the upper end of chamber B opened into a small chamber which had no other entrance. One side of it was completely destroyed. On the remaining slabs were represented the siege and sack of a city standing between two rivers in the midst of groves of palm trees. From the absence of mountains, the nature of the trees and the two rivers, it may be conjectured that the sculptures in this chamber recorded the conquest of some part of Mesopotamia or the subjugation of a people inhabiting that country who had rebelled against the authority of the Assyrian king. The king was represented several times in his chariot amidst groves of palm trees and preceded and followed by warriors. The besiegers were seen cutting down the palms to open and clear the approaches to the city.

From Layard,
Nineveh and Its Remains

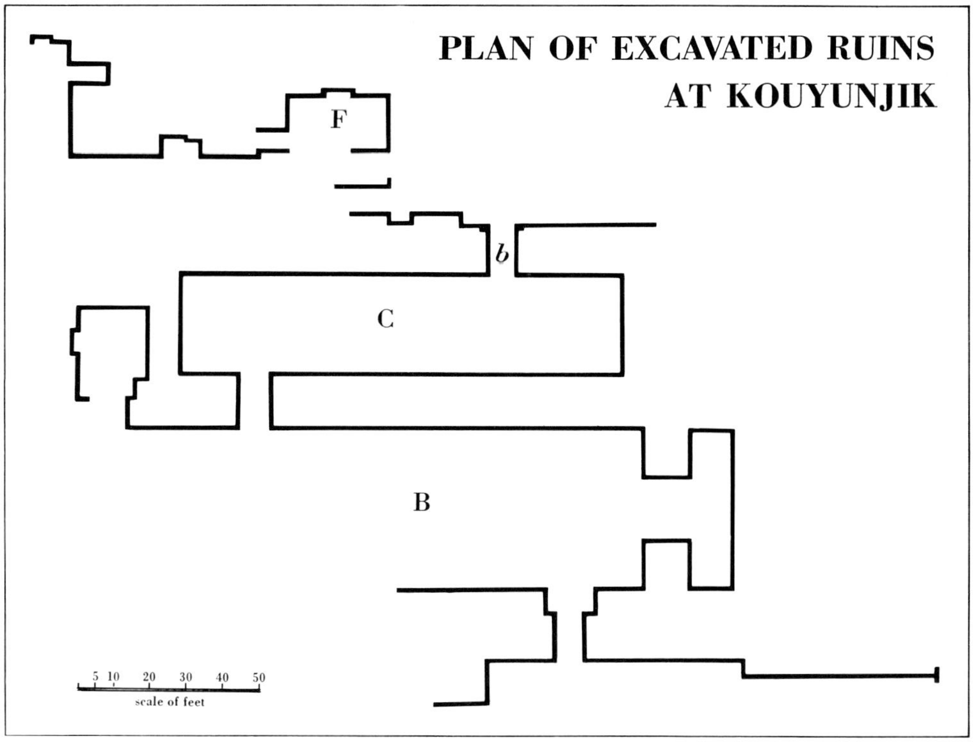

This was the extent of my discoveries at Kouyunjik. The ruins were evidently those of a palace of great extent and magnificence. From the size of the slabs and the number of the figures, the walls, when entire and painted, as they no doubt originally were, must have been of considerable beauty, and the dimensions of the chambers must have added greatly to the general effect. At that time the palace rose above the river, which swept round the foot of the mound. Then also the edifice, now covered by the village of Nebbi Yunus, stood entire above the stream, and the whole quadrangle was surrounded by lofty walls cased with stone, their towers adorned with sculptured alabaster and their gateways formed by colossal bulls. The position of the ruins proves that at one time this was one of the most important parts of Nineveh, and the magnificence of the remains shows that the edifices must have been founded by one of the greatest of the Assyrian monarchs.

My labors in Assyria had now drawn to a close. The funds assigned to the Trustees of the British Museum for the excavations had been expended, and from the instructions sent to me, further researches were not, for the present at least, contemplated. It now, therefore, only remained for me to wind up my affairs in Mosul, to bid adieu to my friends there and to turn my steps homeward, after an absence of some years. The ruins of Nimroud had been again covered up, and its palaces were once more hidden from the eye. The sculptures taken from them had been safely removed to Busrah and were now awaiting their final transport to England. The inscriptions, which promise to instruct us in the history and civilization of one of the most ancient and illustrious nations of the earth, had been carefully copied. On looking back upon the few months that I had passed in Assyria, I could not but feel some satisfaction at the result of my labors. Scarcely a year before, with the exception of the ruins of Khorsabad, not one Assyrian monument was known. Almost sufficient materials had now been obtained to enable us to restore much of the lost history of the country and to confirm the vague traditions of the learning and civilization of its people, hitherto treated as fabulous. It was consequently just at the right moment that they were disinterred, and we have been fortunate enough to acquire the most convincing and lasting evidence of that magnificence and power which made Nineveh the wonder of the ancient world and her fall the theme of the prophets as the most signal instance of divine vengeance. Without the evidence that these monuments afford, we might almost have doubted that the great Nineveh ever existed, so completely "has she became a desolation and a waste."

Before my departure I was desirous of giving a last entertainment to my workmen and to those who had kindly aided me in my labors. On the western side of Kouyunjik there is a small village. This village was chosen for the festivities, and tents for the accommodation of all the guests were pitched around it. Large platters filled with boiled rice and divers inexplicable messes, only appreciated by Arabs and those who have lived with them—the chief components being garlic and sour milk—were placed before the various groups of men and women, who squatted in circles on the ground. Dances were then commenced and were carried on through the greater part of the night. The dancers were happy and enthusiastic and kept up a constant shouting.

At the conclusion of the entertainment I spoke a few words to the workmen, inviting any who had been wronged or ill used to come forward and receive such redress as it was in my power to afford and expressing my satisfaction at the successful termination of our labors without a single accident. One Sheikh Khalaf, a very worthy man who was usually the spokesman on such occasions, answered for his companions. They had lived, he said, under my shadow, and, God be praised, no one had cause to complain. Now that I was leaving they should leave also and seek the distant banks of the Khabour, where at least they would be far from the authorities and be able to enjoy the little they had saved. All they wanted was a note for each man certifying that he had been in my service. This would not only be some protection to them but they would show my writing to their children and would tell them of the days they had passed at Nimroud. I gave a few presents to the principal workmen and their wives, and all were highly satisfied with their treatment.

A few days afterward the preparations for my departure were complete. I paid my last visit to Essad Pasha, called upon the principal people of the town and on the twenty-fourth of June was ready to leave Mosul.

I was accompanied on my journey to Constantinople by Mr. Hormuzd Rassam, Ibrahim Agha and the Bairakdar, and by several members of the household of the late Pasha who were ready, in return for their own food and that of their horses, to serve me on the road. We were joined by many other travelers who had been waiting for an opportunity to travel to the north in company with a sufficiently strong party. The country was at this time very insecure. The failure of the crops had brought parties of Arabs abroad, and scarcely a day passed without the plunder of a caravan and the murder of travelers. The Pasha sent a body of irregular horse to accompany me as far as the Turkish camp which I wished to visit on my way. With this escort, and with my own party, all well armed and prepared to defend themselves, I had no cause to apprehend any accident by the way.

Mr. and Mrs. Rassam, all the European residents and many of the principal Christian gentlemen of Mosul rode out with me to some distance from the town. The greater part of the Arabs insisted upon walking as far as Tel Kef with me. In this village supper had been prepared for the party. We sat on the housetop till midnight. The horses were then loaded and saddled. I bid a last farewell to my Arabs and started on the first stage of our long journey to Constantinople.

Chapter 9

■ I have endeavored to describe the manner in which excavations were carried on among the ruins of Nineveh and the discoveries to which they gave rise.

Three spots have been examined, Nimroud, Kouyunjik and Khorsabad, and of the three, Khorsabad, the smallest, alone thoroughly. Unfortunately, in the Assyrian edifices little but the sculptured slabs has been preserved. To these may be added a few relics such as seals and cylinders, and one or two inscriptions on stones, bricks and tiles. Still the sculptures do furnish us with very interesting and important details, both with regard to the arts and to the manners of the Assyrians. It may not be uninteresting to inquire how Nineveh was built and what knowledge the Assyrians possessed of the science of architecture.

The architecture of a people must naturally depend upon the materials afforded by the country and upon the objects of their buildings. Had the Assyrians, so fertile in invention, so skillful in the arts and so ambitious of great works, dwelt in a country as rich in stone and costly granites and marbles as Egypt or India, it can scarcely be doubted that they would have equaled, if not excelled, the inhabitants of those countries in the magnitude of their pyramids and in the magnificence and symmetry of their rock temples and palaces. But their principal settlements were in the alluvial plains watered by the Tigris and Euphrates. On the banks of those great rivers, which spread fertility through the land and afford the means of easy and expeditious intercourse between distant provinces, they founded their first cities. On all sides they had vast plains, unbroken by a single eminence until they approached the foot of the Armenian hills.

The earliest habitations were probably one story in height. In this respect the dwelling of the ruler scarcely differed from the meanest hut. It soon became necessary, however, that the temples of the gods and the palaces of the kings, depositories at the same time of the national records, should be more conspicuous than the humble edifices by which they were surrounded. The means of defense also required that the castle, the place of refuge for the inhabitants in times of danger or the permanent residence of the garrison, should be raised above the city and should be built so as to afford the best means of resistance to an enemy. As there were no natural eminences in the country, the inhabitants were compelled to construct artificial mounds. Hence the origin of those vast, solid structures, which have defied the hand of time

The Metropolitan Museum of Art,
Gift of John D. Rockefeller, Jr., 1932 (detail)

and, with their grass-covered summits and furrowed sides, rise like natural hills in the Assyrian plains.

The materials for building were at hand, and their preparation required neither much labor nor ingenuity. The soil, an alluvial deposit, was rich and tenacious. The builders moistened it with water and, adding a little chopped straw that it might be more firmly bound together, they formed it into squares which, when dried by the heat of the sun, served them as bricks. In that climate the process required but two or three days. Such were the earliest building materials, and they are used to this day almost exclusively in the same country. We have an allusion to this mode of brick-making in Exodus, for the Egyptians, to harass their Jewish captives, withheld the straw without which their bricks could not preserve their form and consistency.

Huts for the people were speedily raised, the branches and boughs of trees from the banks of the river serving for a roof.

They now sought to build a place of refuge in case of attack or a habitation for their leader or a temple to their gods. It was first necessary to form a platform of sun-dried bricks, thirty or forty feet high, and upon it they erected the royal, or sacred edifice.

The earliest edifices of this nature appear to have been at the same time public monuments in which were preserved the records or archives of the nation, carved on stone. In them were represented in sculpture the exploits of the kings or the forms of the divinities, while the history of the people and invocations to their gods were also inscribed in written characters upon the walls. It was necessary, therefore, to use some material upon which figures and inscriptions could be carved. The plains of Mesopotamia as well as the lowlands between the Tigris and the hill country abound in a kind of coarse alabaster or gypsum. It is easily worked, and its color and transparent appearance are agreeable to the eye. This alabaster, therefore, cut into large slabs, was used in the public buildings.

The walls of the chambers, from five to fifteen feet thick, were first constructed of sun-dried bricks. The alabaster slabs were used as panels. They were placed upright against the walls, care being first taken to cut on the back of each an inscription recording the name, title and descent of the kings undertaking the work. They were kept in their places and held together by iron, copper or wooden clamps and plugs.

The principal entrances to the chambers were formed by gigantic winged bulls and lions with human heads. The smaller doorways were guarded by colossal figures of divinities or priests. No remains of doors or gates were discovered, nor of hinges, but it is probable that the entrances were provided with them. Under the pavement slabs at the entrances were deposited small figures of the gods, probably as a protection to the building. Sometimes, as in the early edifices, tablets containing the name and title of the king, as a record of the time of the erection of the building, were buried in the walls or under the pavement.

The roof was probably formed by beams supported entirely by the walls; smaller beams, planks or branches of trees were laid across them, and the whole was plastered on the outside with mud. Such are the roofs in modern Arab cities of Assyria.

It has been seen that the principle of the arch was known to the Assyrians, a small vaulted chamber of baked bricks having been found at Nimroud; but there are no traces of an arch or vault used on a large scale.

If daylight were admitted into the Assyrian palaces, it could only have entered from the roof. There are no communications between the inner rooms except by the doorways, consequently they could only receive light from above. Even in the chambers next to the outer walls there are no traces of windows.

The walls of the Assyrian cities, as we learn from the united testimony of ancient authors, were of extraordinary size and height. The remains which still exist exceeded in thickness any modern walls. The materials were generally bricks of clay, dried in the sun.

At certain distances in the walls there were gates, sometimes flanked by towers adorned with sculptures, and sometimes formed by gigantic figures such as the winged bulls and lions.

The bas-reliefs and sculptures of the Assyrians, except, probably, those in black marble and basalt, were either partly or entirely painted. Some bricks from Nimroud appear to have been enameled, the colors having been laid on very thickly when in a liquid state and then exposed to the action of fire. Although limited in the number of their colors, the Assyrians displayed considerable taste and skill in their arrangement. The use of a strong black outline is a peculiar feature in Assyrian, as in Egyptian painting. On the colossal lions and bulls forming the entrances, color only remained in the eyes, the pupils having been painted black and the rest filled with a thick white pigment.

Chapter 10

■ The Assyrians were celebrated at a very early period for the magnificence and luxury of their apparel. The robes, as portrayed in the sculptures, confirm the traditions of their beauty and costliness. The dress of the king consisted of a long flowing garment descending to the ankles, elaborately embroidered and edged with fringes and tassels. It was confined at the waist by a girdle to which were attached cords with large tassels falling down almost to the feet. Over this robe a second, nearly of the same length but open in front, appears to have been thrown. It was also embroidered and edged with tassels. On his head he wore a high miter or tiara of peculiar shape, reserved for the monarch alone. Only one band passed round the headdress of the earliest monuments; at a later period two or more were introduced, and the miter itself was higher and more richly ornamented. These peculiarities mark distinctly the respective ages of sculptures in which the figure of the king occurs.

More than one necklace of elegant form was generally suspended round the neck. The arms were encircled by armlets and the wrists by bracelets, all equally remarkable for the taste and beauty of the design and workmanship. The clasps were in the shape of the heads of lions and other animals, and in the center of the bracelets were stars or rosettes which were probably inlaid with precious stones. Earrings of many kinds were worn.

In the shape and workmanship of their arms the Assyrians displayed considerable taste and skill. The king, even in time of peace, appears always to have carried a sword and two daggers. He is sometimes represented leaning, during the celebration of religious ceremonies, upon a bow and holding two arrows in one hand. When returning from war, he frequently raises the two arrows, and this action appears to indicate triumph over his enemies. When not engaged in battle, he is usually portrayed at Khorsabad, and sometimes at Nimroud, with a long staff in his right hand, the other resting on the hilt of his sword. When in battle his arms were the bow and the sword; in the sculptures he is seen using both.

The sandals worn by the king and by his principal officers were formed of a sole either of wood or thick leather to which was attached an upper case covering the heel and side of the foot, but leaving exposed the instep and toes. It was fastened by bands attached to loops and carried twice over the instep. They crossed on the top of the foot and were passed round the great toe and

British Museum (detail)

between it and the adjoining toe.

The sandal represented at Khorsabad, and in sculptures of the same period, is altogether of a different shape. It appears to have consisted of a simple leather covering for the heel held by three strings passing over the instep. It was painted in the bas-reliefs in alternate stripes of red and blue.

The attendants upon the king, both in time of peace and war, were chiefly eunuchs; these persons rose to the highest rank and were not mere servants. In the sculptures eunuchs are represented as commanding in war, fighting both in chariots and on horseback and receiving the prisoners and the heads of the slain after battle. They were also employed as scribes and are seen writing down the number of the heads and the amount of the spoil obtained from the enemy. The countenance of the eunuchs is strongly contrasted in the sculptures with that of the man, and the rounded form, the bloated cheek and double beardless chin at once mark them and distinguish them from females.

Their dress consisted of a long tunic descending to the ankles, resembling the king's in shape and in the richness and elegance of its embroideries. In battle they were armed with the bow and in peace, as well as in war, generally carried a sword and daggers. When represented as attending upon the king, they usually bear a quiver, bow and mace, all probably for his use. At other times they raise a parasol or fan over his head or present him with the sacred cup, on which occasion they are frequently unarmed.

The vizier or prime minister, the principal officers and attendants of the king, were clothed in robes resembling those of the eunuchs. They were armed with swords and daggers and also wore necklaces, bracelets and earrings. A fillet or band, either plain or richly ornamented, frequently confined their hair and encircled their temples.

The Assyrians, besides wearing the numerous ornaments described, most carefully and elaborately plaited their hair and beards. The hair was parted over the forehead and fell from behind the ears on the shoulders in a large bunch of ringlets. The beard was allowed to grow to its full length and, descending low on the breast, was divided into two or three rows of curls. The mustache was also carefully trimmed and curled at the ends.

The costume of the warriors differed according to their rank and the nature of the service they had to perform. Those who fought in chariots and held the shield for the defense of the king are generally seen in coats of scale armor which descend either to the knees or to the ankles. The armor was

British Museum

frequently embossed with groups of figures and fanciful ornaments.

The pointed helmet in the bas-reliefs from the earliest palace of Nimroud appears to have been the most ancient, and in the most general use; it is, indeed, characteristic of the Assyrian warrior. Several were discovered in the ruins: they were of iron, and the rings which ornament the lower part and end in a semicircle in front, were inlaid with copper. Circular iron caps, fitting closely to the head, were also in use at an early period. The helmets of the later monuments of Nimroud, and of those of Khorsabad and Kouyunjik, are frequently surmounted by a curved crest or by a kind of plume.

The arms of the early Assyrians were the spear, the bow, the sword and the dagger. The bows were of two kinds: one long and slightly curved, the other short and almost angular. The two appear to have been carried at the same time by those who fought in chariots.

The arrows were probably made of reeds and were kept in a quiver slung over the back. The king, however, and the great officers of state were followed by attendants who carried the quivers and supplied their masters with arrows. The bow was drawn to the cheek or to the ear and not to the breast. The barbs were of iron and copper. When in battle, it was customary for the archer to hold two arrows in reserve in his right hand; they were placed between the fingers and did not interfere with the motion of the arm drawing the bow. When marching he usually carried the larger bow over his shoulders, having first passed his head through it. The bow of the king was borne by an attendant. The smaller bows were frequently placed in the quiver, particularly by those who fought in chariots. A leather or linen guard was fastened by straps to the inside of the left arm to protect it when the arrow was discharged. The swords were worn on the left side and suspended by belts passing over the shoulders or round the middle; some were short and others long.

The dagger appears to have been carried by all, both in time of peace and war; even the priests and divinities are represented with them. They were worn indifferently on the left and right side or perhaps on both at the same time. Generally two, or sometimes three, were inserted into one sheath, which was passed through the girdle. The handles were most elaborately adorned and were frequently in the shape of the head of a ram, bull or horse, being made of ivory or rare stones.

The spear of the Assyrian footman was short, scarcely exceeding the height of a man; that of the horseman appears to have been considerably longer.

The mace, a weapon in very general use among them, consisted of a short handle, probably of wood, to which was fixed a head, evidently of metal, in the shape of a flower, rosette, lion or bull. To the end of the handle was attached a thong, apparently of leather, through which the hand was passed. I have not found any representation of warriors using the hatchet except when cutting down trees to clear the country preparatory to a siege. It is, however, generally seen among the weapons of those who fought in chariots and was carried in the quiver, with the arrows and short angular bow.

In the bas-reliefs of Kouyunjik, slingers are frequently represented among the Assyrian troops. The sling appears to have consisted of a doubled rope with a thong, probably of leather, to receive the stone; it was swung round the head. The slinger held a second stone in his left hand and at his feet is generally seen a heap of pebbles ready for use.

The shields of the Assyrians were of various forms and materials. In the more ancient bas-reliefs a circular buckler, either of hide or metal, perhaps in some instances of gold and silver, is most frequently introduced. It was held by a handle fixed to the center. Light oblong shields of wickerwork, carried in a similar manner, are also found in the early sculptures, but those of a circular form appear to have been generally used by the charioteers.

The shield used in a siege concealed the whole person of the warrior and completely defended him from the arrows of the enemy. It was made either of wickerwork or of hides. It was furnished at the top with a curved point or a square projection, like a roof, at right angles to the body of the shield, which may have served to defend the heads of the combatants against missiles discharged from the walls and towers of a castle.

The archers, whether on foot or in chariots, were accompanied by shield bearers whose office it was to protect them from the shafts of the enemy. Sometimes one shield covered two archers. The shield bearer was usually provided with a sword which he held ready drawn for defense. The king was always attended in his wars by this officer, and even in peace one of his eunuchs usually carried a circular shield for his use.

British Museum

British Museum

A great part of the strength of the Assyrian armies consisted in chariots and horsemen. The chariots appear to have been used by the king and the highest officers of state. The chariot was used during a siege as well as in open battle. When the king in his chariot formed part of a triumphal procession, armed men led the horses. The chariot was also preceded and followed by men on foot.

The Assyrian chariot varied considerably in form at different periods. As represented on the earliest monuments, it is low with the upper part rounded. To each side were fixed two quivers containing arrows, a small crooked bow, a javelin and a hatchet or battle-ax. The pole was sustained by a forked rod fastened to the forepart of the chariot.

Although the yoke was for two horses, three were generally harnessed to the chariot. There is no indication of traces nor can it be ascertained from the sculptures how the third horse was attached. It was probably intended to supply the place of a killed horse and did not draw.

Behind the chariot was suspended a shield with teeth or bosses, and a spear was placed upright in a rest which was usually in the shape of a human head. The warriors stood upright in the chariot, which does not seem to have been furnished with seats.

The later Assyrian chariot was always drawn by two horses, not by three. The later chariots were often completely covered with ornaments; those represented on the earlier monuments had a very elegant molding, or border, round the sides.

The harness and trappings of the horses were extremely rich and elegant. Plumes waved over their heads, or fanciful crests rose gracefully in an arch above the ears and descended in front to the nostrils.

It is probable that the bit, as well as many ornaments of the bridle and trappings, were of gold and other precious materials.

The chariot horses of the later Assyrian period differed entirely in their trappings and harness from those of the earlier. High plumes, generally three in number and rising one above the other, waved over their heads. Frequently an arched crest and clusters of tassels were placed between their ears. Similar tassels fell over their foreheads and hung round their necks. The harness attached to the yoke was more profusely ornamented with rosettes and fringes than that of the earlier Assyrian chariots, but the ornaments showed less variety and taste.

The manes were either allowed to fall loosely on the neck, were plaited or were cut short and stood erect. In the earlier sculptures, the tails of the horses are simply bound in the center with ribands; in the later, the end is plaited and tied up in a bunch.

The horsemen formed a no less important part of the Assyrian army than the charioteers. They were armed with bows or with long spears. They wore short tunics and their legs and feet were bare. When riding without pads or saddles they sat with their knees almost on a level with the horse's back. After the introduction of saddles, their limbs appear to have been more free and they wore greaves or boots, but were unprovided with stirrups.

When an archer on horseback was in battle, his horse was held and guided by a second horseman who rode by his side. He was then able to discharge his arrows freely. On the monuments of Khorsabad and Kouyunjik, the cavalry are usually armed with the spear. When using this weapon they did not require a second horseman to hold the reins.

The riding horses are less richly and profusely adorned than those in harness, the horsemen being probably of inferior rank to those who fought in chariots. The headstall was surmounted by an arched crest, and round the neck was an embroidered collar ending in a rich tassel or bell.

Chapter 11

■ As the Assyrians possessed disciplined and organized troops, it is probable that they were also acquainted, to a certain extent, with military tactics and that their battles were fought upon some kind of system. Their monuments show that among their enemies, also, there were nations not unacquainted with the military science. They had bodies of troops in reserve; they advanced and retreated in rank and performed various maneuvers.

The king himself, attended by his vizier, his eunuchs and principal officers of state,was present in battle, and not only commanded, but took an active part in the affray. The Assyrian king was accompanied to the war, however distant its seat might be, by his wives, his concubines and his children, and by an enormous retinue of servants. Even his nobles were similarly attended. Their couches were of gold and silver, and the hangings of the richest materials. Vessels of the same precious metals were used at their tables; their tents were made of the most costly stuffs and were even adorned with precious stones. They were also accompanied by musicians, who are represented in the sculptures as walking before the warriors on their triumphant return from battle.

The army was followed by a crowd of sutlers, servants and grooms who added to its bulk and acted as an impediment upon its movements and carried ruin and desolation into the countries through which it passed. This multitude could not depend entirely for supplies upon the inhabitants, whom they unmercifully pillaged, but provisions in great abundance, as well as livestock, were carried with them.

The armies were provided with the engines and materials necessary for the siege of the cities they might meet with in their expedition. If any natural obstructions impeded the approach to a castle, such as a forest or a river, they were, if possible, removed. Rivers were turned out of their courses if they impeded the operations of the army, and warriors are frequently represented in the sculptures cutting down trees which surround a hostile city.

The first step on attacking a hostile city was probably to advance the battering ram. If the castle was built, as in the plains of Assyria and Babylonia, upon an artificial eminence, an inclined plane reaching to the summit of the mound was formed of earth, stones or trees, and the besiegers were then able to bring their engines to the foot of the walls. This road was not unfrequently covered with bricks, forming a kind of paved way up which

British Museum (detail)

British Museum

the ponderous machines could be drawn without much difficulty. The approaches not only enabled the besiegers to push their battering rams up to the castle but at the same time to escalade the walls, the summit of which might otherwise have been beyond the reach of their ladders.

The battering rams were of several kinds. Some were joined to movable towers which held warriors and armed men. The whole then formed one great temporary building, the top of which is represented in the sculptures as on a level with the walls, and even turrets, of the besieged city. In some bas-reliefs the battering ram is without wheels; it was then perhaps constructed on the spot and was not intended to be removed.

The artificial tower was usually occupied by two warriors; one discharged his arrows against the besieged, whom he was able from his lofty position to harass more effectually than if he had been below; the other held up a shield for his companion's defense. Warriors are not unfrequently represented as stepping from the machine to the battlements.

Archers on the walls hurled stones from slings and discharged their arrows against the warriors in the artificial towers, while the rest of the besieged were no less active in endeavoring to frustrate the attempts of the assailants to make breaches in their walls. By dropping a double chain or rope from the battlements, they caught the ram and could either destroy its efficacy altogether or break the force of its blows. Those below, however, by placing hooks over the engine and throwing their whole weight upon them, struggled to retain it in its place.

The besieged, if unable to displace the battering ram, sought to destroy it by fire and threw lighted torches, or firebrands, upon it. But water was poured upon the flames through pipes attached to the artificial tower.

Other engines and instruments of war were employed by the besiegers. With a kind of catapult, apparently consisting of a light wooden frame covered with canvas or hides, they threw large stones and darts against the besieged, who in their turn endeavored to set fire to it by torches. A long staff with an iron head resembling a spear was used to force the stones out of the walls. Mines were also opened, and the assailants sought to enter the castle through concealed passages.

Attempts were made to set fire to the gates of the city by placing torches against them, or to break them open with axes.

Mounting to the assault by ladders was constantly practiced and appears to have been the most general mode of attacking a castle, for ladders are

From Layard,
Nineveh and Its Remains

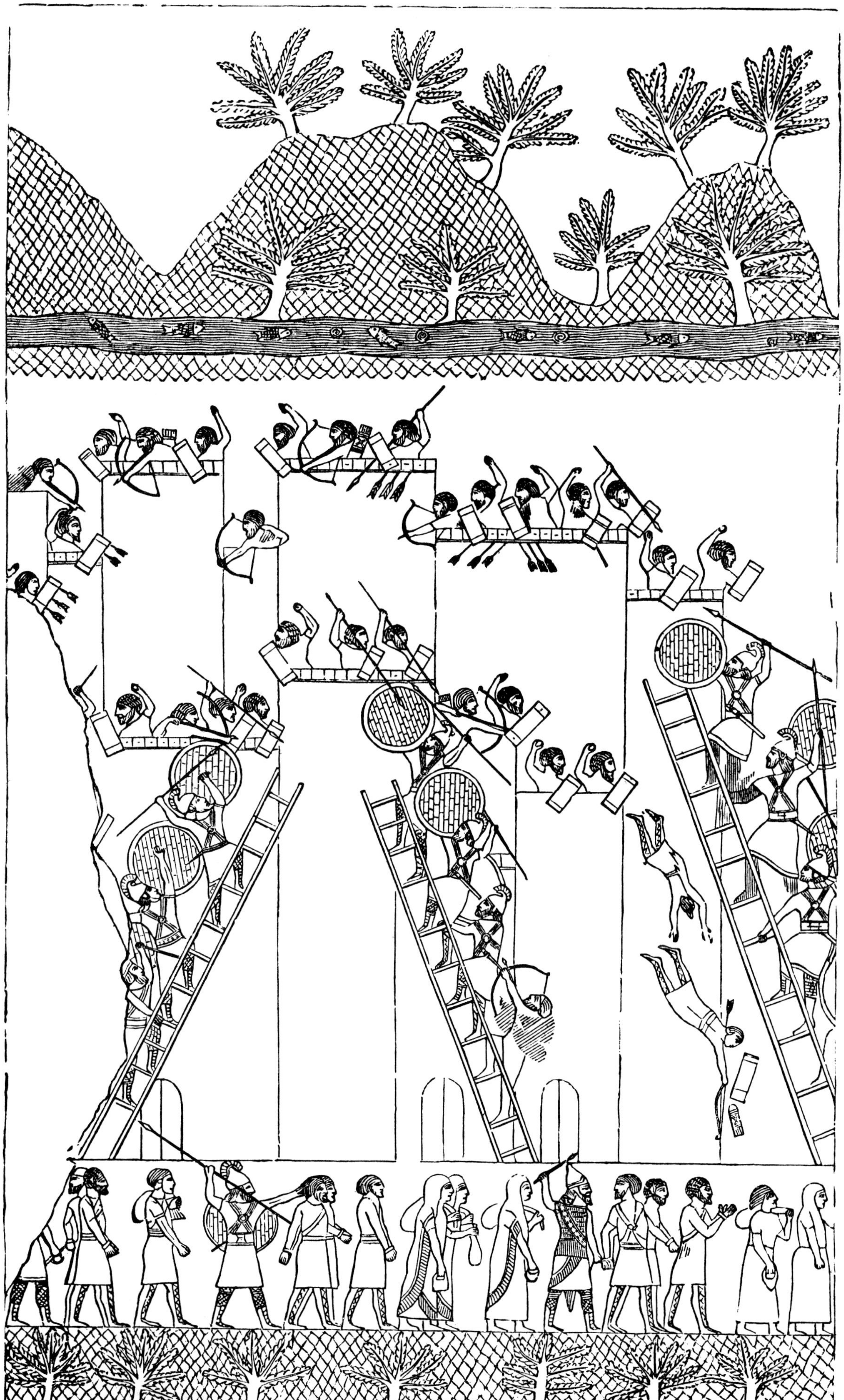

found on those bas-reliefs in which neither the battering ram nor other engines are introduced. They reached to the top of the battlements, and several persons could ascend them at the same time. While warriors armed with the sword and spear scaled the walls, archers posted at the foot of the ladders kept the enemy in check and drove them from the walls.

The besieged manned the battlements with archers and slingers who discharged their missiles against the assailants. Large stones and hot water were also thrown upon those below.

When the battering ram had made a breach and the assault had commenced, the women appeared upon the walls and, tearing their hair or stretching out their hands, entreated for mercy. The men are not unfrequently represented as joining in asking for quarter. When the assailants were once masters of the place, an indiscriminate slaughter appears to have succeeded, and the city was generally given over to the flames. In the bas-reliefs warriors are seen decapitating the conquered and plunging swords or daggers into their hearts, holding them by the hair of their heads. The prisoners were subjected to horrible torments or carried away as slaves.

The women, children and cattle were led away by the conquerors, and it was frequently the custom of the Assyrians to remove the whole population of the conquered country to some distant part of their dominions and to replace it by colonies of their own.

As soon as the soldiers entered the captured city, they began to plunder and then hurried away with the spoil. They led off the horses, carried forth on their shoulders furniture and vessels of gold, silver and other metals and made prisoners of the inhabitants, who probably became the property of those who seized them.

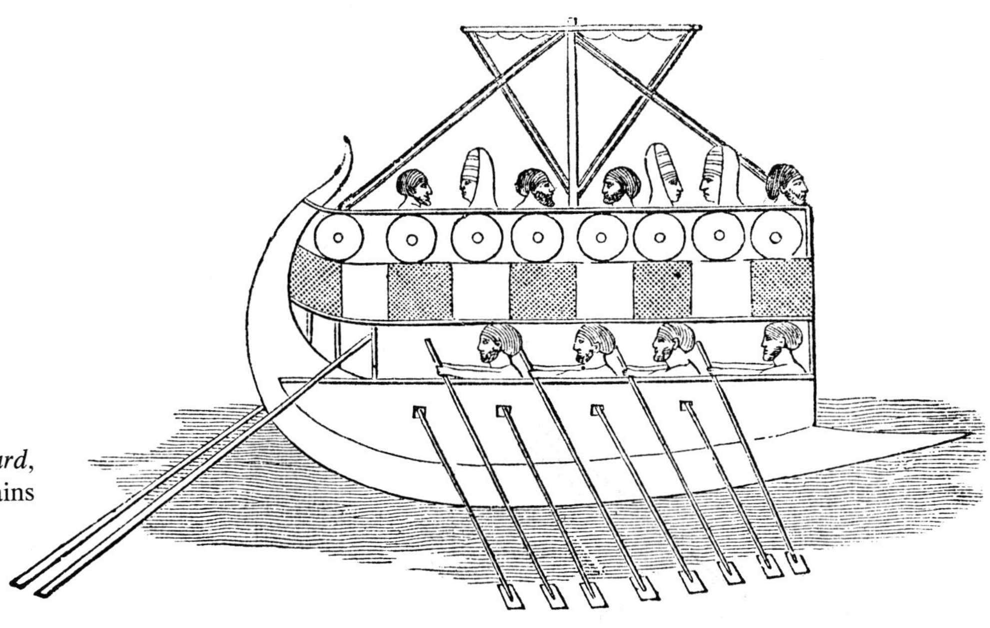

From Layard,
Nineveh and Its Remains

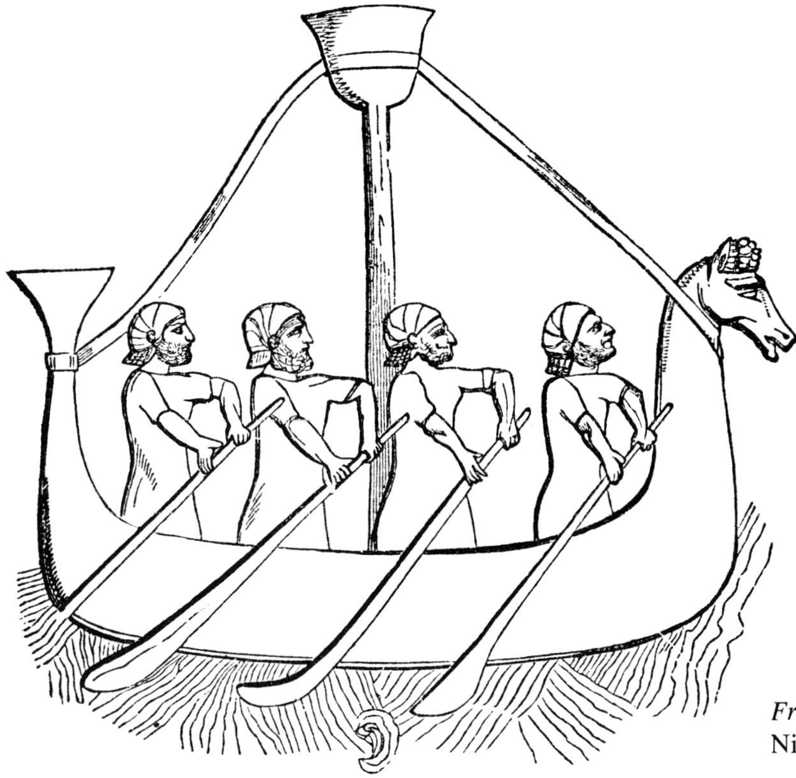

From Layard,
Nineveh and Its Remains

Although the Assyrians were properly an inland people, yet their conquests and expeditions, particularly at a later period, brought them into contact with maritime nations. We consequently find, on the monuments of Khorsabad and Kouyunjik, frequent representations of naval engagements and operations on the seacoast.

The vessels of the Khorsabad sculptures show a considerable advance in the knowledge of shipbuilding. That they did not belong to the Assyrians but to some allied nation appears to be indicated by the peculiar costume of the figures in them. The form of the vessel is that of a sea monster—the prow being in the shape of the head of a horse and the stern in that of the tail of a fish. Several men stand at the oars. The mast, supported by two ropes, appears to be surmounted by a bos, or what is technically called a crow's-nest. But it was in the sculptures of Kouyunjik that vessels were found represented in the greatest perfection.

The forepart [of the larger galleys] rose perpendicularly from a low sharp prow which resembled a plowshare, and being of iron or some other metal, may have been intended, like that of the Roman galley, to sink or disable the enemy's ships. The stern was curved from the keel and rose high above the upper deck. There were two tiers of rowers. Above the rowers was a deck on which stood the armed men. These vessels had only one mast, to the top of which was attached a very long yard held by ropes. In the sculptures, the sails were represented as furled. The number of rowers in the bas-reliefs was generally eight on a side. Only the heads of the upper tier of men were visible; the lower tier was completely concealed, the oars passing through small apertures, or portholes, in the sides of the vessel. Around the sides of the vessels were also suspended the shields of the warriors.

Besides the vessel I have described, a smaller is represented in the same bas-reliefs. It has a double tier of rowers, but the head and stern are differently constructed from those of the larger galley, both being of the same shape and not to be distinguished. They rise high above the water and are flat at the top, with a beak projecting outward. This vessel had no mast and was impelled entirely by oars. On the upper deck are seen warriors armed with spears, and women.

At Kouyunjik, the conquest of a different people appears to have been recorded on the walls of each chamber. Almost the whole of western Asia was overrun by the Assyrians, and their victorious armies, having subdued Syria and Judea and carried away captive their monarchs and their inhabitants, penetrated through Egypt into Ethiopia and Libya. Records of these conquests still exist in Syria, in Cyprus and in various parts of Asia Minor.

Herodotus [the Greek historian who lived in the 5th century B.C.] alludes to the extreme fertility of Assyria and to its rich harvests of corn, the seed producing, according to his account, two or three hundredfold. The blades of wheat and barley grew to full four fingers in breadth. The Assyrians used machines for raising water from the river or from artificial canals when it could not be led into the fields through common conduits. They were generally obliged to have recourse to this artificial mode of irrigation, as the banks of the rivers and consequently those of the canals were high above the level of the water, except during the spring. At that season of the year the streams, swollen by the melting of the snows in the Armenian hills or by violent rains, overflowed their beds.

The only representation of an agricultural instrument yet found in Assyria or Babylonia is that of a plow on a black stone from the ruins opposite Mosul. The plow somewhat resembles in shape that now in common use. On the same tablet is an altar or low building, before which stands a priest apparently performing some religious ceremony; near him are the sacred tree, a bull, a heap of corn or a hill, a palm tree and a square instrument with a small circle or wheel at each corner.

Sesame, millet and corn formed anciently the principal agricultural produce of Assyria.

The frequent representation of hunting scenes in Assyrian sculptures is a proof of the high estimation in which the chase was held by the people. A conqueror and the founder of an empire was, at the same time, a great hunter. His courage, wisdom and dexterity were as much shown in encounters with wild beasts as in martial exploits; he rendered equal services to his subjects whether he cleared the country of wild beasts or repulsed an enemy. We have seen that the Babylonians as well as the Assyrians ornamented the walls of their temples and palaces with pictures and sculptures representing the chase, and that similar subjects were introduced even in the embroideries on their garments. The Assyrians were probably also the inventors of the parks or paradises. In these spacious preserves various kinds of wild animals were continually kept for the diversion of the king and for those who were privileged to join with him in the chase. They contained lions, tigers, wild boars, antelopes and many varieties of birds.

Assyrian youths were probably trained to hunting at an early age. In the Nimroud bas-reliefs, the king is always represented as accompanied in the chase by warriors fully equipped, hunting being the truest method of practicing all such things as relate to war.

On the obelisk are representations of several animals evidently brought from distant countries and presented to the Assyrian king as objects of tribute. The presence of the two-humped camel proves that they came from the East and not from Africa. The small ears of the elephant, on the same obelisk, show that the animal is of the Indian and not of the African species. The rhinoceros on the obelisk is in several respects incorrectly delineated, the sculptor having given it hoofs, a mane on the neck and long hair which appears to have been artificially curled like that of the sacred bull. Specimens of this animal were probably rare in Assyria, and the sculptor may have drawn it from recollection or only from the description of those who had seen it.

The two animals accompanying the rhinoceros are probably an Indian bull and a kind of antelope. The bull has a collar ornamented with tassels round its neck and may have been a sacred animal.

Immediately behind the elephant is a man leading a large monkey without a tail. A man follows with two smaller monkeys, one raising itself on its hind legs, the other sitting on the shoulders of its keeper.

The only birds represented on the Assyrian monuments are the eagle or vulture, the ostrich and the partridge and a few smaller birds at Khorsabad.

The vulture or eagle—for the bird is rarely delineated with sufficient accuracy to enable us to decide which—is continually seen over the heads of the conquerors in battle and in triumphal processions, and was probably typical of victory. It is also represented feeding on the bodies of the slain and flying away with the entrails.

In sea and river scenes fish and shells are introduced. In the rivers are seen crabs, eels or water snakes and small turtles. When the sculptor wished to indicate the sea, he made these fish larger and added others which are only inhabitants of salt water, such as the starfish. A kind of crocodile is also represented in the sea pieces.

With the exception of the vine, palm and fir, the trees of the Assyrian bas-reliefs are conventional in their forms. The sculptor introduced them merely to show the nature of the country in which the events recorded took place.

On the earliest Assyrian monuments one of the most prominent sacred types is the eagle-headed, or vulture-headed, human figure. Not only is it found in colossal proportions on the walls, or guarding the portals of the chambers, but it is also constantly represented among the groups on the embroidered robes. When thus introduced it is generally seen contending with other mythic animals, such as the human-headed lion or bull, and in these contests it appears to be always the conqueror.

The winged human-headed lions and bulls, those magnificent forms which guarded the portals of the Assyrian temples, next deserve notice. Not only are they found as separate sculptures, but, like the eagle-headed figures, are constantly introduced into the groups embroidered on the robes. I have already ventured to suggest the idea which these singular forms were intended to convey—the union of the greatest intellectual and physical powers.

To all these images some meaning was undoubtedly attached. They were emblematical, representing either the attributes of the Deity or certain physical phenomena in nature. But I cannot venture, at present, to conjecture the signification of any of them, nor am I able to determine the character of the winged human figures which so frequently occur on the walls of Assyrian buildings. They may be the representations of presiding deities or genii, or of priests who, during the celebration of sacred ceremonies, assumed that which was believed to be the outward form of the divinities. But such subjects are at present out of my province. I shall be well satisfied, and my literary labors will be amply rewarded, if I have succeeded in an attempt to add a page to the history of mankind by restoring a part of the lost annals of Assyria.

The Metropolitan Museum of Art,
Gift of John D. Rockefeller, Jr., 1932